Improving Teaching and Learning in Science and Mathematics
David F. Treagust, Reinders Duit, and Barry J. Fraser, Editors

Reforming Mathematics Education in America's Cities:
The Urban Mathematics Collaborative Project
Norman L. Webb and Thomas A. Romberg, Editors

What Children Bring to Light:
A Constructivist Perspective on Children's Learning in Science
Bonnie L. Shapiro

STS Education: International Perspectives on Reform
Joan Solomon and Glen Aikenhead, Editors

Reforming Science Education:
Social Perspectives and Personal Reflections
Rodger W. Bybee

Improving Teaching and Learning in Science and Mathematics

EDITED BY

David F. Treagust
Reinders Duit
Barry J. Fraser

Teachers College
Columbia University
New York and London

Published by Teachers College Press, 1234 Amsterdam Avenue, New York, NY 10027

Library of Congress Cataloging-in-Publication Data

Improving teaching and learning in science and mathematics/edited by
    David F. Treagust, Reinders Duit, Barry J. Fraser.
        p.   cm.—(Ways of knowing in science series)
    Includes bibliographical references and index.
    ISBN 0-8077-3480-2 (cloth: alk. paper).—ISBN 0-8077-3479-9
(pbk.: alk. paper)
        1. Science—Study and teaching—United States.   2. Mathematics—
    Study and teaching—United States.   3. Constructivism (Education)—
    United States.   I. Treagust, David F.   II. Duit, Reinders.   III. Fraser, Barry J.
    IV. Series.
    LB1585.3.I55   1996
    372.3'5'044—dc20   95-38578

ISBN 0-8077-3479-9 (paper)
ISBN 0-8077-3480-2 (cloth)

Printed on acid-free paper
Manufactured in the United States of America

03 02 01 00 99 98 97 96   8 7 6 5 4 3 2 1

Contents

List of Figures and Tables

FIGURES

TABLES

Acknowledgement

We would like to thank Di Youdell for her word-processing expertise and dedication in bringing the manuscript to completion.

CHAPTER 1

Overview: Research on Students' Preinstructional Conceptions— The Driving Force for Improving Teaching and Learning in Science and Mathematics

David F. Treagust
Curtin University of Technology, Perth, Australia

Reinders Duit
University of Kiel, Germany

Barry J. Fraser
Curtin University of Technology, Perth, Australia

During the past two decades, research on the role of students' preinstructional conceptions has been influenced by an old pedagogical maxim of Ausubel (1968): "The most important single factor influencing learning is what the learner already knows. Ascertain this and teach . . . accordingly" (p. v). However, the insight expressed in this dictum is not new; it guided the work of educators, including science and mathematics educators, long before Ausubel. For example, the German educator Diesterweg (1790–1866), in his *Guide for German Teachers* (1835/1850), pointed out that it is of utmost importance to start instruction from the student's point of view. He added that it is necessary therefore for the teacher to investigate students' preinstructional conceptions (Jung, 1993). Similarly, the hallmark of Dewey's approach to education was the attempt to connect subject-matter knowledge with the child's experiences. These approaches are in sharp contrast to traditional approaches that often start from facts and concepts outside the child's range of experiences (Prawat, 1992).

Research on students' preconceptions in science started at the end of the nineteenth century with investigations of children's ideas of natural phenomena such as heat, frost, and fire (Hall & Brown, 1903). Some 50 years later, Oakes (1947) reviewed several hundred papers, including the landmark writings of Piaget, dealing with research on students' ideas about natural phenomena. In the middle of the 1970s, there began a boom of research on students' conceptions in science and mathematics that continues today. This research focuses on investigating students' conceptions of science and mathematics principles and concepts such as heat, energy, photosynthesis, fractions, decimals, functions, and shape. In science, 66 percent of the studies available fall into the domain of physics, with biology and chemistry sharing the remaining 34 percent (Duit, 1993). Reviews of relevant studies are provided by Confrey (1990), Driver (1995), Driver and Erickson (1983), Duit (1993), and Wandersee, Mintes, and Novak (1993), whereas bibliographies devoted to students' alternative frameworks in science are provided by Carmichael, Driver, Holding, Phillips, Twigger, and Watts (1990), and Pfundt and Duit (1994). For the different fields of mathematics, no similar bibliographic overviews are available.

The many empirical studies provide ample evidence that students hold preinstructional conceptions in many fields and that these are substantially different from the scientific and mathematical concepts taught in school. Most of these conceptions are held strongly and hence are very resistant to change. Research shows that students learn mathematics and science concepts and principles only to a limited degree, that sometimes students persist almost totally with their preinstructional conceptions, and that sometimes students try to hold on to two inconsistent approaches—one intuitive and one formal. Frequently, students' alternative conceptions are unrecognized by teachers and can affect instruction in unpredicted ways.

In the mid-1970s, science and mathematics educators researched students' conceptions in isolation, topic by topic. When this led to limited success in modifying students' understandings and beliefs, researchers extended the scope of their investigations to include students' and teachers' beliefs about the overall scientific or mathematical enterprise, as well as their beliefs about learning and teaching. For example, philosophy of science research (Carey, Evans, Honda, Jay, & Unger, 1989; Lederman, 1992; Lederman & O'Malley, 1990) has shown that students and teachers often hold rather limited views of the nature of science and mathematics. Many students and teachers are naive realists in that they view science and mathematics knowledge as a faithful copy of the "world outside" and not as tentative human construction. Hence, teachers and students are less likely to engage in teaching and learning approaches that are based on the idea of knowledge as a tentative human construction and that usually are referred to by the term "constructivist" as discussed below in more detail.

Students' views of the teaching and learning process are important for learning. Many students hold a passive view and conceptualize learning as the transfer of canonical knowledge that then is stored in memory. According to this view, science is learned primarily as an accumulation of facts and mathematics as a memorization of formulas. This view of learning as a passive process influences the students' conceptions of what counts as legitimate work in school. Classroom discussions of alternative viewpoints and negotiated consensus are not considered a part of the "work" of the classroom, and simply are viewed as wasted time that hinders efficient progress (Baird & Mitchell, 1986).

Now that research on students' conceptions has been placed within a larger framework, including beliefs of students and teachers about knowledge, learning, and teaching, the research field can inform practice about how to engage in more productive mathematics and science education. This larger framework forms the constructivist perspective of this book. The authors of the chapters that make up this volume subscribe to many shared principles of constructivism but these often are interpreted in independent ways. In the book's three parts, the authors illustrate, respectively, constructivist principles in investigating student understanding (Chapters 2–6), in teaching and curriculum development (Chapters 7–14), and in teacher education (Chapters 15–19).

In this introductory chapter, we discuss briefly the basic ideas of constructivism (and refer to more extensive reviews) and provide a cursory discussion of the challenges and debates concerning constructivism. In our opinion, some of the strengths of constructivism are its potential for informing practice and the variety of ways in which it can be interpreted. After describing the significance of the constructivist view for research and classroom practice, this chapter concludes with an overview of the chapters in this volume.

THE BASIC IDEAS OF CONSTRUCTIVISM

The constructivist view has become the leading theoretical position in education and has become a most powerful driving force in science and mathematics education, particularly during the past decade (Ernest, 1993; Malone & Taylor, 1993; Steffe & Gale, 1995; Tobin, 1993). The appeal of the constructivist view is that it provides a plausible, functional framework for understanding and interpreting experiences of learning and teaching; in this way, constructivism acts as a powerful theoretical referent "to build a classroom that maximises student learning" (Tobin & Tippins, 1993, p. 7). Constructivism is described as consisting of two basic principles, one psychological and the second epistemological, and emphasizes that knowledge cannot be separated from knowing. These two principles, which make up von Glasersfeld's (1990) radical constructivism, include many facets of Piaget's genetic epistemology, and often serve as a reference position for discussions of constructivism in education.

The first principle states that knowledge is not received passively but is built up by the cognizing subject. In other words, it is not possible to transfer ideas into students' heads intact; rather, students construct their own meanings from the words or visual images they hear or see. Consequently, when engaging in this construction of meaning, what the learner already knows is of central importance. This principle is becoming generally accepted by science and mathematics educators.

The second principle states that the function of cognition is adaptive and enables the learner to construct viable explanations of experiences. Consequently, knowledge about the "world outside" is viewed as a human construction. A "reality" outside is not denied, but it is possible to know about that reality only in a personal and subjective way (von Glasersfeld, 1990). This second principle is controversial for some educators as, at first sight, it appears that it might never be possible to understand what anyone else is saying or meaning, and it might not be possible to understand the meaning that is given to what we all accept as known, particularly in science and mathematics. Rather than being concerned with knowledge as the representation of the truth, constructivism focuses on the way in which learners construct viable and useful knowledge. According to this position, the only constructions that survive are those that prove to be successful in dealing with the multiple contexts in which the learner is engaged. Von Glasersfeld (1993) introduced the term "radical constructivism" to contrast it with constructivism mainly or only built on prior knowledge. Such a position often is referred to as weak or trivial constructivism, the latter term having a pejorative connotation.

Researchers into constructivism believe that knowledge not only is personally constructed but that it also is socially mediated (Prawat, 1993; Taylor, 1993). This third aspect highlights the fact that, although individuals have to construct their own meaning of a new phenomenon or idea, the process of constructing meaning always is embedded within the social setting of which the individual is a part. Social constructivism in mathematics education is described by Bauersfeld (1988, 1994), Cobb,

Wood, and Yackel (1991), and Confrey (1995), while social constructivism in science education is described by Cobern (1993), Solomon (1989), and Tobin (1993).

In the introductory chapter to Part II, Duit and Confrey describe five assumptions shared by constructivist science and mathematics educators. However, it must be emphasized that there is no "constructivist teaching and learning," but simply teaching and learning. According to a constructivist view, students actively construct their own knowledge whenever they learn something. However, this occurs even in settings that are not informed by constructivist ideas, that is, in strictly traditional settings in which learners are assumed to receive knowledge passively (Millar, 1989). The major difference is that constructivist teaching and learning approaches explicitly aim to help students to make the constructions that lead to understanding of the scientific or mathematical points of view. This approach is a delicate balance of students' own activities and guidance by the teacher, or by a teaching medium such as can be provided by computers. From our perspective, it is absurd to expect students to be able to construct science and mathematics conceptions without any guidance on the basis of their preexisting conceptions alone.

CHALLENGES TO AND DEBATES ABOUT THE CONSTRUCTIVIST VIEW

If constructivist approaches are to be implemented effectively, teachers' traditional beliefs about transmission approaches to teaching and absorptionist views of learning need to be challenged. Prawat (1992) describes four beliefs that serve as impediments to the constructivist view of teaching and learning in that they underlie traditional, transmissionist approaches to teaching. First, many teachers view the learner and the content as separate and static entities that must be reconciled. The second set of beliefs, which Prawat terms "naive constructivism," concerns the tendency to equate activity with learning. The third set of beliefs, which Prawat challenges on the basis of limited research evidence, is the distinction between comprehension and application giving rise to the idea that learning is hierarchical and that generalization leads to transfer. The fourth set of beliefs is that the curriculum is a fixed entity consisting of well-ordered content mastered according to predetermined criteria. Certainly, these four sets of beliefs appear to be consistent with constraints coming from the power structures of the school system that work against changes.

Constructivism is open to different interpretations. But, regardless of the range of different views about constructivism (see, for example, Good & Schlagel, 1992; Prawat, 1993), there is no doubt that the constructivist view has been a productive influence in research on students' and teachers' conceptions and in teaching and learning science and mathematics (Steffe & Gale, 1995). Even critics of the constructivist view (e.g., Matthews, 1993; Solomon, 1994) admit that this perspective inspired researchers and curriculum developers.

As might be expected of such a broadly based epistemological conception, the critics of constructivism take different positions, though they tend to be centered around four issues: Constructivism is simply common sense; it has epistemological flaws; it leads to the denial of the existence of the physical world; and excessive focus on the individual does not take social issues into account.

According to Strike (1987), the two principles of constructivism are simply common sense, that is, they are shared by almost everyone who deals with science and mathematics education. Similarly, Suchting (1992) claims that the weak terminology of radical constructivism allows trivialization of the main ideas as principles that are shared by almost everyone, like the insight that there is a certain aspect of human construction in all scientific knowledge. Solomon (1994) takes a position similar to those of Strike and Suchting, arguing that "constructivism only arrived on the science education scene once it acquired a new vocabulary to match new intentions" (p. 3). In making such claims, however, these authors appear to ignore the large body of research that shows that students and teachers often hold limited views of the philosophy and nature of science (Lederman, 1992) and that, in the learning process, metacognition, for example, is difficult to achieve (White & Mitchell, 1994). Further findings illustrate that teachers who expressed constructivist views in interviews did not necessarily implement these principles in classroom situations (Fischler, 1994), or had difficulties implementing these principles even when they had strong intentions (Tobin, 1992). Consequently, from our perspective, the statement that these principles are accepted widely and hence simply are common sense should be regarded with caution; principles might be accepted in theory but very often they are not put into practice.

A second point of criticism concerns a certain "empiricist" nature of radical constructivism (Matthews, 1993; Suchting, 1992). However, the claim, in constructivism as well as in empiricism, that experiences are the key source of learning is not sufficient. In constructivism, unlike empiricism, new knowledge does not come from experiences alone. Sense impressions, for instance, have to pass the filter of prior and preinstructional conceptions in order to make sense to the observer. What appears to be argued appropriately in Matthews's (1993) discussion of the relationship between empiricism and constructivism is the overemphasis on the individual in both positions. Further elaboration of this epistemology is needed, not so much to defend this point of view as to save the power that this epistemology undoubtedly has for learning science and mathematics.

The third critique, that radical constructivism leads to the denial of the existence of a physical outside world, is incorrect given that von Glasersfeld (1992) states that radical constructivism is ontologically neutral. As described earlier, radical constructivism is consistent with the idea of a real existing world outside; all it denies is the possibility of any certain knowledge of that reality.

Other critics of radical constructivism focus on its neglect of social influences on the inner subjective world of the individual and on the individual's construction processes (Ernest, 1993; O'Loughlin, 1992). However, von Glasersfeld (1995) acknowledges the importance of the social context of which the individual learners are part when they construct their knowledge, although the social issue is not addressed further in his radical constructivism.

In general, it would appear to be a healthy state of affairs that a view of learning comprising both psychological and epistemological principles is so well embraced by teachers and researchers and also is criticized by a cadre of colleagues, mainly of a philosophical inclination, who believe that constructivism is not necessarily a perfect answer to improving teaching and learning.

SIGNIFICANCE OF THE CONSTRUCTIVIST VIEW FOR RESEARCH AND CLASSROOM PRACTICE

Even acknowledging the critiques summarized above, the constructivist view still not only allows a surprisingly consistent explanation of major empirical findings concerning the significance of students' preinstructional conceptions in the learning processes, but also facilitates predictions of outcomes of instruction. Despite the need for a more consistent understanding of the implications of constructivism in science and mathematics education, the contemporary view already provides a powerful basis for understanding learning and teaching. For example, in Part I of this volume, insightful approaches for investigating students' understanding of science and mathematics concepts are described.

The constructivist view allows us to explain that students' preinstructional conceptions are difficult to change by instruction because they guide or even determine students' sense-making process when information is provided by the teacher or the textbook. The chapters in Part II of this volume describe promising new approaches to teaching and curriculum development that are based on this insight. The teacher has to become a facilitator of each individual's personal and social construction processes by establishing classroom interaction patterns that promote each individual's exploration and resolution of ideas within the social-cultural context. Teaching and learning roles are characterized by negotiation rather than imposition, and problem solving is achieved through conceptual understanding rather than the application of a prescribed method (Cobb, 1988).

Likewise, these same principles of constructivism have been used in teacher education in developing the interesting and powerful approaches described in each of the five chapters in Part III of this volume. Furthermore, there have been many attempts to inform teachers about the results of research into students' conceptions and about a new view of teaching and learning. But much more has to be done to incorporate constructivist principles into many more science and mathematics classrooms. Nevertheless, the constructivist view (whether under this label or another) is likely to continue to be a most powerful and influential perspective.

The issue of the social environment and the individual should be given considerable attention, including the balance between self-development and guidance. Constructivist approaches should not focus as much on self-development as appears to have been the case so far. The interplay of conceptions of content and metacognitive conceptions of the learning process should be researched further. Whenever science or mathematics content is taught, something about the conceptions of science or mathematics (that is, something of a philosophy of science or mathematics) is taught implicitly. Similarly, when a philosophy of science or mathematics is addressed explicitly, the content can be made more understandable to students.

SCOPE OF BOOK

The underlying theme of this book illustrates how constructivist ideas can be used by science and mathematics educators for research and the further improvement of educational practice. In this book, authors from various parts of the world describe their work investigating students' conceptions, improving teaching and curricula, and enhancing teacher education in science and mathematics contexts.

The book comprises an introduction and three subsequent parts. This introductory chapter explains how the content of the book is informed by a critical perspective of constructivism and provides an overview of the organization of the book. Each part commences with an introductory major chapter providing a broad review of the field, and subsequent chapters deal with specific issues. Part I, "Investigating Student Understanding of Science and Mathematics," provides information about how to investigate students' understanding of content and concepts in science and mathematics. Part II, "Improving Curriculum and Teaching in Science and Mathematics," describes studies of successful attempts to improve the teaching and learning of science and mathematics by reorganizing the curriculum and through associated changes in teaching and learning methods. Part III, "Implementing Teacher Change in Science and Mathematics," examines analytical and holistic approaches to teacher education, and provides an overview of recent developments in this field that are informed by constructivism.

Part I: Investigating Student Understanding of Science and Mathematics

In the introductory chapter, "Investigating Student Understanding as a Prerequisite to Improving Teaching and Learning in Science and Mathematics" (Chapter 2), Duit, Treagust, and Mansfield describe issues that must be considered when using any method to investigate students' conceptual understanding in science and mathematics, and review a variety of these methods. In Chapter 3, "Concept Mapping: A Tool for Improving Science Teaching and Learning," Novak outlines a broad range of issues involved in the use of concept maps to improve student understanding of science content and reports research showing that the use of concept maps can be a powerful tool for designing, organizing, and revising lectures, textbooks, and curricula. In Chapter 4, "Interviews About Instances and Interviews About Events," Carr provides the rationale for the effective use of an interview technique in which student understanding of concepts is ascertained through their responses to instances or noninstances of a particular concept or phenomenon. In Chapter 5, "Computer-Video–Based Tasks for Assessing Understanding and Facilitating Learning in Geometrical Optics," Goldberg and Bendall present their approach for examining student conceptual difficulties in geometrical optics using an instructional strategy that incorporates a videodisc system to assess understanding and encourage learning. In Chapter 6, "Using Teaching Experiments to Enhance Understanding of Students' Mathematics," Steffe and D'Ambrosio describe a teaching experiment designed to provide understanding of students' construction of the concept of multiplication.

Part II: Improving Curriculum and Teaching in Science and Mathematics

In the introductory chapter to Part II, "Reorganizing the Curriculum and Teaching to Improve Learning in Science and Mathematics" (Chapter 7), Duit and Confrey review valid or stable principles of learning that have significance for science and mathematics education and have been used to improve curriculum and teaching in science and mathematics. The other seven chapters in this part present approaches, based on one or more of these principles, in different contexts. While each of the approaches is different, they share much in common in their applications of constructivist principles in the classroom. In Chapter 8, "Curriculum Development as Research: A Constructivist Approach to Science Curriculum Development and Teaching," Driver and Scott discuss an approach to curriculum development in which groups of teachers become involved in implementing curriculum change and assessing student learning in their school. In Chapter 9, "Strategies for Remediating Learning Difficulties in Chemistry," Ben-Zvi and Hofstein describe how each teaching process in a specific context can be used to bridge the gap between the student and the complex and abstract nature of chemistry. In Chapter 10, "Using Student Conceptions of Parallel Lines to Plan a Teaching Program," Mansfield and Happs describe how a program of teaching materials resulted in long-term conceptual understanding of aspects of parallel lines among a large sample of eighth-grade students. In Chapter 11, "Teaching for Conceptual Change," Hewson outlines a model of learning as conceptual change, identifies two general guidelines of this model for teaching, and considers the factors necessary to support it. In Chapter 12, "Contrastive Teaching: A Strategy to Promote Qualitative Conceptual Understanding of Science," Schecker and Niedderer describe a case study of contrastive teaching in upper-school mechanics that illustrates how this approach can be used to enable students to differentiate similar concepts in the same domain. In Chapter 13, "Concept Substitution: A Strategy for Promoting Conceptual Change," Grayson describes a teaching approach with poorly prepared university science students that involves the substitution of a new concept for a student's scientifically acceptable idea that previously was given an incorrect name and identified as a misconception. In Chapter 14, "Changing the Curriculum to Improve Student Understandings of Function," Confrey and Doerr describe an innovative approach to curriculum change involving the teaching and learning of functions.

Part III: Implementing Teacher Change in Science and Mathematics

In the introductory chapter to the third part of the book, entitled "Analytical and Holistic Approaches to Research on Teacher Education"

(Chapter 15), Tobin examines attempts to improve teacher education from analytical and holistic perspectives. The other four chapters in this part provide specific instances of reform in teacher education which have been informed by constructivist principles. In Chapter 16, "Metacognitive Strategies in the Classroom," Baird and White describe a project in which teachers and students were encouraged to use metacognitive strategies to improve the overall performance of the students. In Chapter 17, "A Constructivist Perspective on Science Teacher Education," Northfield, Gunstone, and Erickson describe programs aimed at developing a constructivist philosophy in student teachers of science. In Chapter 18, "Diagnosis of Teachers' Knowledge Bases and Teaching Roles When Implementing Constructivist Teaching/Learning Approaches," Hand describes the development and implementation of a model for diagnosing constructivist teaching of science and mathematics. In Chapter 19, "Implementing Teacher Change at the School Level," Gallagher describes how teacher cooperation with university personnel and school administrators enabled science and mathematics teachers to improve their own teaching and that of their colleagues.

ACKNOWLEDGMENTS

We would like to acknowledge the critical comments on an earlier version of this chapter made by Peter Taylor (Curtin University of Technology, Australia), Diane Grayson (University of Natal, South Africa), and Jere Confrey (Cornell University, United States).

REFERENCES

Ausubel, D. P. (1968). *Educational psychology: A cognitive view.* New York: Holt, Rinehart and Winston.

Baird, J. R., & Mitchell, I. J. (1986). *Improving the quality of teaching and learning—An Australian case study.* Melbourne, Victoria: Monash University.

Bauersfeld, H. (1988). Interaction, construction, and knowledge: Alternative perspectives for mathematics education. In T. Cooney & D. Grouws (Eds.), *Effective mathematics teaching* (pp. 27–42). Reston, VA: National Council of Teachers of Mathematics.

Bauersfeld, H. (1994). Theoretical perspectives on interaction in the mathematics classroom. In R. Biehler, R. Scholz, R. Straesser, & B. Winkelmann (Eds.), *Didactics of mathematics as a scientific discipline* (pp. 133–146). Dordrecht, The Netherlands: Kluwer.

Carey, S., Evans, R., Honda, M., Jay, E., & Unger, C. (1989). "An experiment is when you try it and see if it works": A study of grade 7 students' understanding of the construction of scientific knowledge. *International Journal of Science Education, 11,* 514–529.

Carmichael, P., Driver, R., Holding, B., Phillips, I., Twigger, D., & Watts, M. (1990). *Research on students' conceptions in science—A bibliography.* Leeds, England: University of Leeds.

Cobb, P. (1988). The tension between theories of learning and instruction in mathematics education. *Educational Psychologist, 23,* 87–103.

Cobb, P., Wood, T., & Yackel, E. (1991). A constructivist approach to second grade mathematics. In E. von Glasersfeld (Ed.), *Radical constructivism in mathematics* (pp. 157–176). Dordrecht, The Netherlands: Kluwer.

Cobern, W. W. (1993). Contextual constructivism: The impact of culture on the learning and teaching of science. In K. Tobin (Ed.), *The practice of constructivism in science education* (pp. 51–69). Washington, DC: American Association for the Advancement of Science.

Confrey, J. (1990). A review of the research on student conceptions in mathematics, science and programming. In C. Cazden (Ed.), *Review of research in education, 16* (pp. 3–56). Washington, DC: American Educational Research Association.

Confrey, J. (1995). The relationships between radical constructivism and social constructivism. In L. Steffe & J. Gale (Eds.), *Constructivism in education* (pp. 185–225), Hillsdale, NJ: Erlbaum.

Diesterweg, F. A. W. (1835/1850). *Wegweiser zur Bildung für Deutsche Lehrer (Guide for German Teachers).* Essen, Germany: Bädecker.

Driver, R. (1995). Constructivist approaches to science teaching. In L. Steffe and J. Gale (Eds.), *Constructivism in education* (pp. 385–400), Hillsdale, NJ: Erlbaum.

Driver, R., & Erickson, G. L. (1983). Theories-in-action: Some theoretical and empirical issues in the study of students' conceptual frameworks in science. *Studies in Science Education, 10,* 37–60.

Duit, R. (1993). Research on students' conceptions—Developments and trends. In J. Novak (Ed.), *Proceedings of the Third International Seminar on Misconceptions and Educational Strategies in Science and Mathematics Education* [Electronic media]. Ithaca, NY: Cornell University.

Ernest, P. (1993). Constructivism, the psychology of learning, and the nature of mathematics: Some critical issues. *Science & Education, 2,* 87–93.

Fischler, H. (1994). Concerning the difference between intention and action: Teachers' conceptions and actions in physics teaching. In I. Carlgren, G. Handal, & S. Vaage (Eds.), *Teachers' minds and actions: Research on teachers' thinking and practice* (pp. 165–180), London, England: Falmer Press.

Good, R., & Schlagel, R. (1992). Contextual realism in science and science teaching. In S. Hills (Ed.), *The history and philosophy of science in science education: Proceedings of the International Conference on the History and Philosophy of Science and Science Teaching* (Vol. 1, pp. 415–422). Kingston, Ontario: Faculty of Education, Queen's University.

Hall, G. S., & Brown, C. E. (1903). Children's ideas of fire, heat, frost and cold. *Pedagogic Seminar, 10,* 27–85.

Jung, W. (1993). Uses of cognitive science to science education. *Science & Education, 2,* 31–56.

Lederman, N. G. (1992). Students' and teachers' conceptions of the nature of science: A review of research. *Journal of Research in Science Teaching, 29,* 331–360.

Lederman, N. G., & O'Malley, M. (1990). Students' perceptions of tentativeness in science: Development, use, and sources of change. *Science Education, 74,* 225–239.

Malone, J. A., & Taylor, P. C. S. (Eds.). (1993). *Proceedings of Topic Group 10 at the Seventh International Congress on Mathematical Education: Constructivist interpretations of teaching and learning mathematics.* Perth, Australia: Curtin University of Technology.

Matthews, M. (1993). Constructivism and science education: Some epistemological problems. *Journal of Science Education and Technology, 1,* 359–369.

Millar, R. (1989). Constructive criticisms. In R. Driver (Ed.), *International Journal of Science Education* [Special issue], *11,* 587–596.

Oakes, M. E. (1947). *Children's explanations of natural phenomena.* New York: Teachers College Press.

O'Loughlin, M. (1992). Rethinking science education: Beyond Piagetian constructivism toward a sociocultural model of teaching and learning. *Journal of Research in Science Teaching, 29,* 791–820.

Pfundt, H., & Duit, R. (1994). *Bibliography—Students' alternative frameworks and science education.* Kiel, Germany: Institute for Science Education (IPN), University of Kiel.

Prawat, R. S. (1992). Teachers' beliefs about teaching and learning: A constructivist perspective. *American Journal of Education, 100,* 354–395.

Prawat, R. S. (1993). The value of ideas: Problems versus possibilities in learning. *Educational Researcher, 22*(6), 5–12.

Solomon, J. (1989). Social influence or cognitive growth. In P. Adey (Ed.), *Adolescent development and school science* (pp. 195–199). London, England: Falmer Press.

Solomon, J. (1994). The rise and fall of constructivism. *Studies in Science Education, 23,* 1–19.

Steffe, L., & Gale, J. (Eds.). (1995). *Constructivism in education.* Hillsdale, NJ: Erlbaum.

Strike, K. (1987). Towards a coherent constructivism. In J. Novak (Ed.), *Proceedings of the Third International Seminar on Misconceptions and Educational Strategies in Science and Mathematics Education* (pp. 67–78). Ithaca, NY: Cornell University.

Suchting, W. A. (1992). Constructivism deconstructed. *Science & Education, 1,* 223–254.

Taylor, P. C. S. (1993). Collaborating to reconstruct teaching: The influence of researcher beliefs. In K. Tobin (Ed.), *The practice of constructivism in science education* (pp. 267–298). Washington, DC: American Association for the Advancement of Science.

Tobin, K. (1992). Referents for making sense of science teaching. *International Journal of Science Education, 15,* 241–254.

Tobin, K. (Ed.). (1993). *The practice of constructivism in science education.* Washington, DC: American Association for the Advancement of Science.

Tobin, K., & Tippins, D. (1993). Constructivism as a referent for teaching and learning. In K. Tobin (Ed.), *The practice of constructivism in science education* (pp. 3–21). Washington, DC: American Association for the Advancement of Science.

von Glasersfeld, E. (1990). An exposition on constructivism: Why some like it radical. In R. Davis, C. Maher, & N. Noddings (Eds.), *Constructivist views on the teaching and learning of mathematics* (Journal of Research in Mathematics Education Monograph Number 4, pp. 19–29). Reston, VA: National Council of Teachers of Mathematics.

von Glasersfeld, E. (1992). A constructivist's view of learning and teaching. In R. Duit, F. Goldberg, & H. Niedderer (Eds.), *Research in physics learning: Theoretical issues and empirical studies* (pp. 29–39). Kiel, Germany: Institute for Science Education (IPN), University of Kiel.

von Glasersfeld, E. (1993). Questions and answers about radical constructivism. In K. Tobin (Ed.), *The practice of constructivism in science education* (pp. 23–38). Washington, D.C.: American Association for the Advancement of Science.

von Glasersfeld, E. (1995). A constructivist approach to teaching. In L. Steffe & J. Gale (Eds.), *Constructivism in education* (pp. 3–15), Hillsdale, NJ: Erlbaum.

Wandersee, J. H., Mintes, J. J., & Novak, J. D. (1993). Research on alternative conceptions in science. In D. Gabel (Ed.), *Handbook of research on science teaching* (pp. 177–210). New York: Macmillan.

White, R. T., & Mitchell, I. J. (1994). Metacognition and the quality of learning. *Studies in Science Education, 23,* 21–37.

PART I

Investigating Student Understanding of Science and Mathematics

CHAPTER 2

Investigating Student Understanding as a Prerequisite to Improving Teaching and Learning in Science and Mathematics

Reinders Duit
University of Kiel, Germany

David F. Treagust
Curtin University of Technology, Perth, Australia

Helen Mansfield
University of Hawaii, Honolulu, United States

Investigating students' conceptions not only reveals important insights into students' ways of thinking and understanding in science and mathematics, but also can help researchers and teachers revise and develop their own science and mathematics knowledge. The awareness of alternative points of view, namely, the students' perspectives, allows researchers and teachers to see their own views in totally new ways that can result in major reconstruction of their science and mathematics knowledge or their convictions about how this knowledge should be presented in the classroom. In this chapter, we first describe issues that must be considered when using any method to investigate students' conceptual understanding. Only by taking these issues into account can we be sure that the information gained from these probes can be credible and useful for improving science and mathematics teaching and learning. Second, we review those methods that appear to be of most value for investigating students' conceptual understanding of science and mathematics content topics such as sound propagation, chemical bonding, evolution, geometry, and multiplication. Our view of investigating student understanding is in accord with what White and Gunstone (1992) call "probing under-standing" as a means to improve classroom teaching.

ISSUES IN INVESTIGATING UNDERSTANDING

This section considers the following issues that must be taken into account when investigating student understanding: an appreciation of the interpretation process; interpretation strategies; the symmetrical rela-tionships between researcher and research subject; whether conceptions are brought to light or created; and the investigation itself as a learning process.

The Interpretation Process: Understanding Understanding

At the heart of constructivism is the idea that every process of inter-pretation is a construction based on the conceptions already held by the researcher. Attempts at understanding students' understanding are de-termined fundamentally by these constructions so that every interpreta-tion of another person's understanding is always a tentative hypothesis and never can lead to objective truth about that person's understanding. What researchers call "students' conceptions" are actually their own con-ceptions of students' conceptions. Much care is necessary in planning, carrying out research, and especially in interpretation, in order to remain sensitive to one's own conceptions, ideas, beliefs, and prejudices about others' conceptions.

Many cases in the literature on students' conceptions illustrate that it is apparent that researcher and student have not understood one another. In interviews, for instance, the researcher tries to make sense of students' answers from his or her own science and mathematics perspective, whereas the students could be arguing on quite different grounds. For example, in interviews concerning the concept of light, Jung (1987a) de-scribes the misunderstandings between the interviewer and student that point to the difficulties of the instant interpretation processes that are necessary in interview situations. If nothing, or if only a little, is known about students' conceptions in the field the interview is addressing, then the interviewer's conceptions can limit severely the understanding of stu-dents' understanding achieved by the interview. These same difficulties hold for the interpretation of data from other sources, especially from open-ended questions, when there is the danger of interpreting students' responses in terms of meanings that are far from the students' own ideas. In extreme cases, the interpretation brings to light more of the interpreter's understanding (and misunderstanding) of the topic investigated than of the students' understanding. A spiral-like process appears to operate in which the interpreter first develops ideas of students' understanding, proves these ideas by going back to the data (e.g., an interview), refines the achieved understanding of students' understanding, and so on.

Interpretation Strategies

To address these deficiencies of data collection and interpretation, Jung (1987b) proposed an approach that begins by searching students' responses for comments that appear to be the most informative. From these comments, a first attempt at understanding (Jung speaks about a first theory of students' understanding) evolves and then is developed further in several rounds in the spiral process. In order not to get trapped in one's own ideas, Jung (1987b) proposed the use of deliberately different points of view to interpret interview transcripts. These views might be the commonly accepted science conceptions, ideas stemming from the philosophy and history of science and mathematics, and results from previous research on students' conceptions. Of course, all of these points of view can lead to an interpretation that is quite different from the students' actual conceptions. There appear to be some cases for which parallelisms between students' conceptions and conceptions in the history of science and mathematics are mainly the constructions of interpreters who are very familiar with the history of science or mathematics. However, an interpretation without a previous theory of any kind is impossible because the interpreter will see nothing in the data.

Although it is necessary to use a theory to inform every interpretation, it is valuable to distinguish between guidance by theory and inductive procedures (Jung, 1987b). When using guidance by theory, researchers scan the data for facets that verify the previous theory. On the other hand, inductive procedures examine the data as far as possible with no explicitly fixed theory in mind. Phenomenologists (e.g., Marton, 1981) appear to have a preference for inductive procedures, compared with guidance by theory, in that they try to look at the data with an unprejudiced mind and seem to assume that the data, at least to a certain extent, speak for themselves. In every interpretation process, both strategies play a certain role in that there is a complementarity between verification and induction, with induction helping one avoid being blinded totally by a theory that actually is not well suited to understanding students' understanding.

In conducting research to investigate students' understanding of science and mathematics concepts, raw data, such as students' answers in interviews or free responses on pencil-and-paper tests, are organized by a set of categories that are developed in the spiral interpretation process. The final interpretation has to be made on the basis of the categories and the original transcript so that responses that have been put into a particular category are interpreted in that context (Kesidou & Duit, 1993). Such a contextual interpretation is necessary because the same set of words can have totally different meanings in different contexts.

Symmetrical Relationships Between Researcher and Researched Subject

In traditional research approaches, there is usually a clear distinction between the researcher and the researched person (e.g., the student). The constructivist view, however, demands a certain revision of this framework in that the relationship between researcher and researched subject is moved, at least partly, from the asymmetrical to the symmetrical. Both researcher and researched subject are actors in a communication situation and both try to make sense of the other's understanding. If research includes more persons, the same holds true. Therefore, as the researcher investigates the student's understanding, the student also tries to discover the researcher's understanding. A similar symmetry holds for the role of the teacher in the classroom. Here, it is not only the students who have to be viewed as learning subjects but the teacher, too. This symmetry is highlighted by the key idea that a teacher informed by constructivism is a facilitator of information, thoughts, and ideas, rather than a superimposer or transmitter of knowledge. Similarly, researchers and others taking part in an action research study are viewed as people with equal rights and, in principle, with equal roles.

Bringing Conceptions to Light or Creating Conceptions

If students' conceptions of a specific concept, phenomenon, or principle are investigated, the reports usually speak about the conception students hold about this topic. But it has to be taken into consideration whether each student really does possess these conceptions. Indeed, most methods are not able to give any hint about whether a conception possessed by the student is brought to light or whether the methods created the conception that was observed. In fact, there are good reasons to be careful because if students are asked about concepts, phenomena, or principles they never have seen or heard before, they could invent and create ideas to please the researcher. Students might not answer what they think, but what they think the interviewer expects them to answer (see Carr, Chapter 4, this volume).

As the interviewer probes students' understanding by asking questions about particular relationships or particular aspects of a concept, students could adjust their conceptions to take account of these relationships. Consequently, changes made during an interview are not necessarily permanent; it seems likely that permanent modifications of conceptual frameworks often require the construction of several links between new information and different aspects of existing knowledge and beliefs. However, every interview can be regarded as an opportunity for students to reflect on, reconsider, and reconstruct their conceptions and so cannot be regarded as eliciting students' unchanging views.

Investigating Understanding Viewed as Learning Situations

During the interview situation, students can engage in the process of creating ideas for the first time to make some sense of their intuitive thoughts. That is, the questions posed by the interviewer, in the context suggested by the interviewer, can provide the opportunity or even the necessity for students to investigate their own existing views, to construct or reconstruct new explanations for the views they hold, or to extend existing views or create new ones to account for the structure and subtleties of the context the interviewer is inviting them to consider for the first time. Consequently, every use of a research method should be viewed as a learning situation for the student, and in principle every method can be used as an instructional tool.

Most research studies in the field of students' conceptions follow a snapshot approach by investigating students' understanding at one particular time or by taking one or two further snapshots, for instance before and after instruction. Such an approach is limited because only status reports can be provided and conclusions about student development are possible only with much care. To obtain such information, stroboscopic-like multiple snapshots or learning process studies are necessary. Steffe and D'Ambrosio's method of the teaching experiment (see Chapter 6, this volume), for example, carefully follows the changes in a student's thinking during a sequence of interviews.

METHODS FOR INVESTIGATING STUDENTS' UNDERSTANDING

We have organized methods for investigating students' understanding in terms of whether they involve naturalistic settings, interviews, conceptual relationships, diagnostic test items, and computerized diagnosis. In principle, all these methods are suitable also for investigating students' ideas of the nature of science and mathematics knowledge, and the nature and role of proof in mathematics, or students' views of the learning process.

Naturalistic Settings

Examples of methods for which the researcher observes students in naturalistic learning situations and tries not to influence these situations significantly are observations of classroom activities and of students' verbal and nonverbal actions when dealing with a task. Usually the naturalistic teaching and learning events are audiotaped or videotaped and verbatim transcripts are made of the lesson or activity. Field notes of the lesson or activity taken by the researcher are written up later to provide an intensive account of the observations. Observations in naturalistic settings and classroom observations are used to investigate students' ideas in a content area and to develop new ways of teaching within a constructivist framework (Cobb, Wood, & Yackel, 1990; Confrey, 1990).

Classroom observations and the resulting documentation of classroom discussions have been used early in research on students' conceptions. One of the forerunners of this research field in Germany, Zietz (1937), used protocols he wrote after giving a lesson. He explicitly stated that this method is best suited for investigations of students' conceptions that aim to improve classroom science teaching because it is important for the teacher to get to know the conceptions that students develop in classroom situations. For this reason, the method appears to be most valuable for teachers. Detailed descriptions of lessons not only provide useful information about students' ideas of the concepts taught, but they also make teachers more sensitive to their students' way of thinking in general (Solomon, 1985).

In a number of studies, the focus is on small groups working on a task. Tiberghien (1980) observed a small number of children in naturalistic classroom settings over a period of time when they carried out experiments on heat. Schwedes (1985) reported observations of a small group of students who were allowed to follow their own ideas with only a little guidance given by the teacher when dealing with simple electric circuits over several sessions. Students' actions when building electric circuits were given equal importance in the researcher's interpretations as the students' verbal statements.

Schoenfeld (1983, 1985) explored students' strategies through observation of pairs and small groups of students as they worked together to solve mathematical problems. Students were encouraged to speak aloud as they solved problems, generating protocols for later analysis by the researcher. Balacheff (1991), who also has used interviews with pairs of students to explore their behaviors as they undertook geometric proofs, found that the social context of such interviews was very complex and that usually there were tensions because of the different individual motivations and commitments of the students.

The circumstances in which people generate verbal reports of their thinking affect their problem solutions in a variety of subtle ways. While single-person protocols can be uncontaminated by social concerns, the effects of the environment in which the student is placed could be strongest when the person is working alone, in contrast to working with the support of another student. When a small group is working together, the social dynamics might skew discussions to reflect only one person's point of view or solutions offered by some members of the group might be reinforced or discounted by others, with either positive or negative effect on the solution strategies. Two-person protocols can provide the richest data, because the number of plausible alternative solution paths is restricted and the social dynamics are not as complex as in larger groups. The two-person sessions studied by Schoenfeld (1983) allowed him to observe students' executive decision making and higher-order procedural knowledge, which are key objectives of modern mathematics curricula (Australian Educational Council, 1991; National Council of Teachers of Mathematics, 1989).

A unique chance to observe students' verbal and nonverbal actions in an informal setting is provided by studies in interactive science centers (Beiers & McRobbie, 1992; Feher, 1990). Such studies, which include students' nonverbal actions in dealing with science and mathematics tasks, provide some insight into aspects of students' understanding that are not addressed by studies that take mainly verbal data into account. If verbal and nonverbal actions are both observed, information is provided about the interactions between nonverbal and verbal understandings.

Interviews

Interviews undoubtedly are a very powerful research method for investigating students' understanding of science and mathematics concepts and processes, but they are also very demanding tools. Much experience in interviewing is necessary before the appropriate interview skills can be developed; further, much background knowledge is needed to interpret students' responses. However, interviews used by teachers in classrooms, when there is not the same rigorous interpretation process as is necessary for research purposes, can reveal key issues in students' thinking. In addition, protocols of classroom discussions can help teachers to develop the sensitivity for different students' ideas that is necessary in teaching approaches informed by constructivism.

Interviews can be divided into those conducted with individuals and those with groups, with the former being used much more frequently than the latter. In science education, group interviews with demonstrations have been used, for example, by Gilbert and Pope (1986) to investigate understanding of energy. Treagust and Smith (1989) together interviewed groups of four students to ascertain their understanding of gravity in the solar system. Steffe (1991) interviewed pairs of young children to investigate the role of social interactions in the construction of early number concepts. The strength of group interviews appears to be that the development of ideas in the interaction processes between students can be studied. Group interviews model the elicitation phase that is central in teaching approaches informed by constructivism (see Duit & Confrey, Chapter 7, this volume).

The common types of interviews include clinical Piagetian interviews, interviews about instances, and interviews about events (see Carr, Chapter 4, this volume) and teaching experiments. Thinking-aloud techniques, in which students are asked to think aloud while dealing with a task or solving a problem, can be viewed as interviews, and have been extensively employed in studies of students' problem-solving strategies (Charles & Silver, 1988; Garofalo & Lester, 1985; Larkin & Rainard, 1984; Schoenfeld, 1983, 1985).

Also, it can be advantageous to merge interviewing with questionnaire methods to allow the investigation of conceptions held by large numbers of students. In one strategy, an interview is followed by an experiment for which students write down their predictions and explain them. Subsequently, the experiment is carried out for students to observe and comment on whether their predictions were right. White and Gunstone (1992) refer to such a procedure as Prediction-Observation-Explanation; a specific demonstration is performed by the teacher who asks questions that are similar to an interview.

Piaget's clinical interview has become the prototype for most interview methods because it aims at investigating students' understanding of science and mathematics concepts and phenomena and hence follows and keeps track of students' ideas. Piagetian clinical interviews typically present students with an experiment and, following familiarization with the apparatus, students are asked to explain what will happen if an experiment is carried out (e.g., if a weight at a balance beam is moved to another place). Prediction and actual behavior with the apparatus are compared and form the point of departure for investigating students' ideas; in its original form, the Piagetian interview aimed at assessing the state of cognitive development of students. Piagetian researchers in science and mathematics education in the 1970s and 1980s used interviews to distinguish between concrete operational and formal operational thinking.

In mathematics education research, the expression "teaching experiment" is used to describe studies for which changes are introduced deliberately into the teaching situation by researchers (Koehler & Grouws, 1992, p. 119; Steffe & D'Ambrosio, Chapter 6, this volume). An example of a teaching experiment in science education is described by Lunetta, van den Berg, and Katu (1993). These changes might be instructional activities that are constructed and modified during the experimental period, partly as a result of information gained about the students' knowledge, attitudes, or beliefs during instruction. In teaching experiments, not only is the impact of the researchers on the experimental situation acknowledged, but the interaction between intervention and student characteristics is the focus of study (Cobb et al., 1991).

Conceptual Relationships

Conceptual relationships in science and mathematics can be investigated by pencil-and-paper measures such as concept mapping and relational diagrams, which are described briefly below. Other measures, less well known, include association tasks, tree construction tasks, graph construction, networks, and semantic differentials. Quantitative proximity measures (Shavelson, 1985) have been shown not to be very useful for investigating understanding. Consequently, in science and mathematics education, identification of conceptual relationships has been employed mainly by qualitative methods.

The most useful and best researched of the conceptual relationship measures is the concept map, and this is due primarily to the initial work of Novak from the late 1970s to the present time (see Chapter 3, this volume). Concept maps, developed from Ausubel's assimilation theory of cognitive learning, depict the hierarchy and relationships among concepts and are intended to provide evidence of a student's thinking in that the relationships between concepts are presented clearly and alternative conceptions can be identified easily. Researchers have used concept maps to identify features of students' understanding of various topics in science (e.g., Ross & Munby, 1991) and to trace the development of understanding of various science topics over the course of instruction (e.g., Roth & Roychoudhury, 1993). By means of the computer, which has become a powerful tool to investigate students' understanding, large concept maps can be built up by students and analyzed to identify how concepts are related (Beyerbach & Smith, 1990; Fisher et al., 1990). Concept maps have been used less frequently in mathematics education research, although Hasemann and Mansfield (in press) and Mansfield and Happs (1992) have reported using concept maps with young students and tracing the development of understanding over extended periods of time.

While the value of concept mapping is much claimed in many research studies, researchers must conduct more empirical studies to validate this statement. An example of such research is a meta-analysis (Horton et al., 1993) of 19 concept-mapping studies that met strict criteria of well-defined experimental models with controls. Of these 19 studies, only 2 dealt with younger students, and these were not in science or mathematics. The analysis showed that concept mapping generally has positive effects on both student achievement and attitudes.

White and Gunstone (1992) illustrated the use of relational diagrams or Venn diagrams to probe students' understanding of science and other concepts. A similar method, called "drawing concept circles," has been employed by Wandersee (1987) in biology learning. The advantage of this method is that students' attention can be led to facets that different concepts have in common (i.e., the extent to which there is overlap between classes of objects, events, or abstractions) as well as to the distinctive facets. These diagrams take a short time to complete and can be used with classes of any size and with students at any level from mid-elementary upward; their nonverbal form can expose vagueness in conceptions that are disguised easily in verbal responses.

Diagnostic Test Items

Most research into students' understanding in science and mathematics used interviews or concept maps. Less attention has been paid to conventional test items, whether pencil-and-paper or computer-generated, that are likely to be useful for classroom teachers. However, researchers have used pencil-and-paper measures, such as free-response items, to examine a range of ideas held by large numbers of students, either following in-depth clinical interviews with other students or as a broad investigative measure. The items are designed to address known areas of student conceptions and are similar to those given in interviews about instances and interviews about events. A major limitation of this type of item is the difficulty in interpreting student responses. Generally speaking, if the items have not been thought out carefully and have not been field tested, the results can be very difficult to interpret.

Researchers have used multiple-choice items to examine students' understanding in specified and limited content areas of science and mathematics, as opposed to using such items for assessment purposes. However, this method has been ignored largely because of the strong and negative association between behaviorism and multiple-choice items. Such a concern is justified when authors do not specify the origin of the distracters and when items do not investigate conceptual understanding. In this regard, there have been some improvements made by researchers who have developed distracters for items based on students' answers to essay questions and on other open-ended questions in a limited content domain in secondary school biology (Tamir, 1971) and on students' responses to previous tests in mechanics at the first-year university level (Hestenes, Wells, & Swackhamer, 1992).

Treagust (1995) describes an approach using two tiers of multiple-choice distracters per item to diagnose students' conceptual understanding of specified content areas in science. In these items, the first tier of multiple-choice distracters involves a content response while the second tier involves a reasoning response. The source of the items is specified clearly; the conceptual area is documented by propositional knowledge statements needed to understand the concepts, and the items are developed based on the known literature and responses from students to free-response items and interviews. An essential aspect of this method is the necessity to field-test items so that, as far as possible, they are representative of the range of student responses to the concept being investigated. Using this method, diagnostic test items have been developed in covalent bonding (Peterson, Treagust, & Garnett, 1989) and photosynthesis and respiration in plants (Haslam & Treagust, 1987). These diagnostic tests provide an opportunity for both teachers and researchers to diagnose student learning without the need for interviews.

Computerized Diagnosis

The computer has become a powerful tool for investigating students' understanding. Goldberg and Bendall (see Chapter 5, this volume), for example, designed a computer-aided videodisc system that allows investigation of students' understanding in the field of geometrical optics and guides students toward the science view. The computer has been used as a diagnostic tool to investigate conceptual understanding in genetics (Browning & Lehmann, 1988; Simmons & Kinnear, 1990), heat and temperature (Nachmias, Stavy, & Avrams, 1990), and simple electric circuits (Grob, Pollak, & von Rhoeneck, 1992). Working within Logo environments, researchers have investigated students' levels of geometric thinking (Olson, Kieren, & Ludwig, 1987) and their understanding of various geometric topics such as plane figures (Clements & Battista, 1989), transformations (Olson, Kieren, & Ludwig, 1987), symmetry (Clements & Battista, 1990), and angles (Kull, 1986).

CONCLUSION

In this chapter, we have described several issues that must be considered when using any method to investigate students' conceptual understanding. These issues involve the means and strategies of interpretation, symmetric relationships between researcher and students, and recognizing what is and can be occurring conceptually for a student during a conceptual probe. A complete understanding of these issues is necessary before data emanating from conceptual probes can be deemed credible and useful. Subsequently, we examined several investigative methods described under naturalistic settings, interviews, conceptual relationships, diagnostic test items, and computerized diagnosis, and provided examples from research in various domains of science and mathematics. The outcomes from investigations performed with these methods can provide essential information to improve teaching and learning in science and mathematics education. Several methods for investigating student understanding are described in the following chapters of Part I; how such information can be used is discussed in Part II.

REFERENCES

Australian Education Council. (1991). *A national statement on mathematics for Australian schools*. Melbourne, Australia: Curriculum Corporation.

Balacheff, N. (1991). The benefits and limits of social interaction: The case of mathematical proof. In A. J. Bishop, S. Mellin-Olsen, & J. van Dormolen (Eds.), *Mathematical knowledge: Its growth through teaching* (pp. 175–192). Dordrecht, The Netherlands: Kluwer.

Beiers, R. J., & McRobbie, C. J. (1992). Learning in interactive science centres. *Research in Science Education, 22*, 38–44.

Beyerbach, B. A., & Smith, J. M. (1990). Using a computerized concept mapping program to assess preservice teachers' thinking about effective teaching. *Journal of Research in Science Teaching, 27*, 961–971.

Browning, M. E., & Lehmann, J. D. (1988). Identification of student misconceptions in genetics problem solving via computer programs. *Journal of Research in Science Teaching, 25*, 747–761.

Charles, R., & Silver, E. A. (Eds.). (1988). *The teaching and assessing of mathematical problem solving*. Hillsdale, NJ: Erlbaum.

Clements, D. H., & Battista, M. T. (1989). Learning of geometric concepts in a Logo environment. *Journal for Research in Mathematics Education, 20*, 450–467.

Clements, D. H., & Battista, M. T. (1990). The effects of Logo on children's conceptualizations of angle and polygons. *Journal for Research in Mathematics Education, 21*, 356–371.

Cobb, P., Wood, T., & Yackel, E. (1990). Classrooms as learning environments for teachers and researchers. In R. B. Davis, C. A. Maher, & N. Noddings (Eds.), *Constructivist views on the teaching and learning of mathematics* (Monograph Number 4, pp. 125–146). Reston, VA: National Council of Teachers of Mathematics.

Cobb, P., et al. (1991). Assessment of a problem-centered second-grade mathematics project. *Journal for Research in Mathematics Education, 22*, 3–29.

Confrey, J. (1990). A review of the research on student conceptions in mathematics, science and programming. In C. B. Cazden (Ed.), *Review of research in education* (pp. 3–56). Washington, DC: American Educational Research Association.

Feher, E. (1990). Interactive museum exhibits as tools for learning explorations with light. *International Journal of Science Education, 12*, 35–49.

Fisher, K. M., et al. (1990). Computer-based concept mapping. *Journal of College Science Teaching, 19*, 347–352.

Garofalo, J., & Lester, F. (1985). Metacognition, cognitive monitoring, and mathematical performance. *Journal for Research in Mathematics Education, 16*, 163–176.

Gilbert, J., & Pope, M. (1986). Small group discussions about conceptions in science: A case study. *Research in Science and Technological Education, 4*, 61–76.

Grob, K., Pollak, V., & von Rhoeneck, Ch. (1992). Computerized analysis of students' ability to process information in the area of basic electricity. In R. Duit, F. Goldberg, & H. Niedderer (Eds.), *Research in physics learning: Theoretical issues and empirical studies* (pp. 296–309). Kiel, Germany: Institute for Science Education (IPN), University of Kiel.

Hasemann, K., & Mansfield, H. (in press). Concept mapping in research on mathematical knowledge development: Background, methods, findings and conclusions. *Educational Studies in Mathematics.*

Haslam, F., & Treagust, D. F. (1987). Diagnosing secondary students' misconceptions of photosynthesis and respiration in plants using a two-tier multiple choice instrument. *Journal of Biological Education, 21*, 203–211.

Hestenes, D., Wells, M., & Swackhamer, G. (1992). Force concept inventory. *The Physics Teacher, 30*, 141–158.

Horton, P. B., et al. (1993). An investigation of the effectiveness of concept mapping as an instructional tool. *Science Education, 77*, 95–111.

Jung, W. (1987a). Verstaendnisse und Missverstaendnisse (Understanding and misunderstanding). *Physica Didactica, 14*, 23–30.

Jung, W. (1987b). Understanding students' understandings: The case of elementary optics. In J. D. Novak (Ed.), *Proceedings of the Second International Seminar on Misconceptions and Educational Strategies in Science and Mathematics* (Vol. 3, pp. 268–277). Ithaca, NY: Cornell University.

Kesidou, S., & Duit, R. (1993). Students' conceptions of the second law of thermodynamics—An interpretive study. *Journal of Research in Science Teaching, 30*, 85–106.

Koehler, M. S., & Grouws, D. A. (1992). Mathematics teaching practices and their effects. In D. A. Grouws (Ed.), *Handbook of research on mathematics teaching and learning* (pp. 115–126). New York: Macmillan.

Kull, J. A. (1986). Learning and Logo. In P. F. Campbell & G. G. Fein (Eds.), *Young children and microcomputers* (pp. 103–130). Englewood Cliffs, NJ: Prentice-Hall.

Larkin, J. H., & Rainard, B. (1984). A research methodology for studying how people think. *Journal of Research in Science Teaching, 21,* 235–254.

Lunetta, V. N., van den Berg, D., & Katu, N. (1993). Teaching experiment methodology in the study of electric concepts. In J. Novak (Ed.), *Proceedings of the Third International Seminar on Misconceptions and Educational Strategies in Science and Mathematics* [Electronic media]. Ithaca, NY: Cornell University.

Mansfield, H. M., & Happs, J. C. (1992). Using grade eight students' existing knowledge to teach about parallel lines. *School Science and Mathematics, 92,* 450–454.

Marton, F. (1981). Phenomenography—Describing conceptions of the world around us. *Instructional Science, 10,* 177–200.

Nachmias, R., Stavy, R., & Avrams, R. (1990). A micro-computer based diagnostic system for identifying students' conceptions of heat and temperature. *International Journal of Science Education, 12,* 123–132.

National Council of Teachers of Mathematics. (1989). *Curriculum and evaluation standards for school mathematics.* Reston, VA: Author.

Olson, A. T., Kieren, T. E., & Ludwig, S. (1987). Linking Logo, levels, and language in mathematics. *Educational Studies in Mathematics, 18,* 359–370.

Peterson, R. F., Treagust, D. F., & Garnett, P. J. (1989). Development and application of a diagnostic instrument to evaluate grade 11 and 12 students' concepts of covalent bonding and structure following a course of instruction. *Journal of Research in Science Teaching, 26,* 301–314.

Ross, B., & Munby, H. (1991). Concept mapping and misconceptions: A study of high-school students' understandings of acids and bases. *International Journal of Science Education, 13,* 11–23.

Roth, W.-M., & Roychoudhury, A. (1993). Using vee and concept maps in collaborative settings: Elementary education majors construct meaning in physical science courses. *School Science and Mathematics, 93,* 237–244.

Schoenfeld, A. H. (1983). Episodes and executive decisions in mathematical problem-solving. In R. Lesh & M. Landau (Eds.), *Acquisition of mathematics concepts and processes* (pp. 345–395). San Diego, CA: Academic Press.

Schoenfeld, A. H. (1985). Making sense of "out loud" problem-solving protocols. *The Journal of Mathematical Behavior, 4,* 171–191.

Schwedes, H. (1985). The importance of water circuits in teaching electric circuits. In R. Duit, F. Jung, & Ch. von Rhoeneck (Eds.), *Aspects of understanding electricity* (pp. 319–329). Kiel, Germany: Schmidt & Klaunig.

Shavelson, R. J. (1985, March–April). *The measurement of cognitive structure.* Paper presented at the annual meeting of the American Educational Research Association, Chicago, IL.

Simmons, P. E., & Kinnear, J. F. (1990). A research method using microcomputers to assess conceptual understanding and problem solving. *Research in Science Education, 20,* 263–271.

Solomon, J. (1985). Classroom discussion: A method of research for teachers? *British Educational Research Journal, 11,* 153–162.

Steffe, J. P. (1991). The constructivist teaching experiment: Illustrations and implications. In E. von Glasersfeld (Ed.), *Radical constructivism in mathematics education* (pp. 177–194). Dordrecht, The Netherlands: Kluwer.

Tamir, P. (1971). An alternative approach to the construction of multiple choice test items. *Journal of Biological Education, 5,* 305–307.

Tiberghien, A. (1980). Modes and conditions of learning—an example: The learning of some aspects of the concept of heat. In W. F. Archenhold, R. Driver, A. Orton, and C. Wood-Robinson (Eds.), *Cognitive development research in science and mathematics* (Proceedings of an international seminar, pp. 288–309). Leeds, England: University of Leeds.

Treagust, D. F. (1995). Diagnostic assessment of students' science concepts. In S. Glynn & R. Duit (Eds.), *Learning science in the schools: Research reforming practice* (pp. 327–346). Hillsdale, NJ: Erlbaum.

Treagust, D. F., & Smith, C. L. (1989). Secondary students' understanding of gravity and the motion of planets. *School Science and Mathematics, 89,* 380–391.

Wandersee, J. (1987). Drawing concept circles: A new way to teach and test students. *Science Activities, 24*(4), 9–20.

White, R., & Gunstone, R. (1992). *Probing understanding.* London, England: Falmer Press.

Zietz, K. (1937). Kindliche Erklaerungsversuche fuer Naturerscheinungen (Children's explanations of natural phenomena). *Zeitschrift fuer Paedagogische Psychologie (Journal for Educational Psychology), 38,* 219–228.

CHAPTER 3

Concept Mapping: A Tool for Improving Science Teaching and Learning

Joseph D. Novak
Cornell University, Ithaca, New York, United States

Concept mapping originated in 1972 from a research program that required a way to represent changes in the knowledge structures of students over a 12-year span of schooling (Novak & Musonda, 1991). In this research, students were interviewed periodically from grades 1 through 12 to monitor changes in their conceptual understanding, especially their understanding of the particulate nature of matter. The large accumulation of interview transcripts required a more efficient means of representing knowledge structures and changes in knowledge structures. Consequently, the technique of concept mapping was developed and we have found this to be a useful tool in a variety of applications, including helping students to "learn how to learn" (Novak & Gowin, 1984).

Our version of concept mapping is rooted in a constructivist epistemology that assumes that human beings construct meanings for events and objects that occur in their experience. We define "concept" as a perceived regularity in events or objects designated by a label (usually a word). Concept maps serve to show relationships between concepts, and it is from these relationships that concepts derive their meaning. Figure 3.1 illustrates our view of the nature of concept maps.

Although concept maps remain useful as a research tool to represent knowledge structures, we have also found them to be useful in a variety of applications, including facilitation of meaningful learning, design of instructional materials, identification of misconceptions or alternative conceptions, evaluation of learning, facilitation of cooperative learning, and encouragement of teachers and students to understand the constructed nature of knowledge (Novak, 1990a, 1990b; Novak & Wandersee,

Improving Teaching and Learning in Science and Mathematics. Copyright © 1996 by Teachers College, Columbia University. All rights reserved. ISBN 0-8077-3479-9 (pbk), ISBN 0-8077-3480-2 (cloth). Prior to photocopying items for classroom use, please contact the Copyright Clearance Center, Customer Service, 222 Rosewood Dr., Danvers, MA 01923, USA, tel. (508) 750-8400.

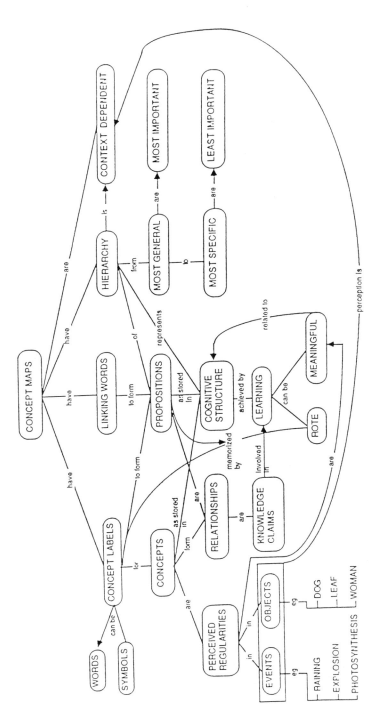

Figure 3.1. Concept Map Showing Key Features of Good Concept Maps and Psychological and Epistemological Ideas Underlying Concept Mapping

1990). In all of these applications, concept mapping serves as a tool for empowerment of both teachers and learners. The fundamental objective is to help individuals make sense out of experience and to construct their own meanings for experience.

The purpose of this chapter is to illustrate the diverse ways in which concept mapping can be used to improve science teaching and learning. In particular, concept mapping is considered in relation to meaningful learning, instructional designs, alternative conceptions, student evaluation, cooperative learning, constructivism, and the empowerment of teachers and students.

ENCOURAGING MEANINGFUL LEARNING

All children begin life as cognitive blank slates, and they must construct a conceptual framework out of experience impinging on them largely through a process of discovery learning during their early years. It now is recognized that, immediately after birth, infants begin to organize their experience and that, by the age of two years, they begin to code these experiences using language. This is an extraordinary process of meaning-making and is done autonomously by children with high success, resulting in the use of language to acquire and transmit meanings with considerable proficiency by the age of three or four years. By school age, children have acquired a substantial vocabulary representing the meanings of several thousand concepts. In short, they have become highly successful *meaningful* learners in their preschool experience.

Unfortunately, most school practice involves arbitrary recitation of definitions or relationships (such as fill-in-the-blank worksheets) that have little or no meaning to the child and result in essentially rote-learning practices. By grade 3 or 4, many students have moved from predominantly high levels of meaningful learning to predominantly rote-mode learning for which the relationships between concepts are arbitrary and are not constructed by the student (Novak & Gowin, 1984). This move toward rote-mode learning in all curricular areas leads to disempowerment of the learner, and this is especially true in science and mathematics, which require large bodies of organized information for understanding. As an illustration, Figure 3.2 provides an example of a concept map constructed by a six-year-old child after 20 minutes of instruction in how to make concept maps. All of the relationships shown on the map are meaningful and are accurate representations of the concept designated by the words. The word "vapor" is missing from this map, because it apparently had no meaning for this child. By contrast, in a concept map

concepts   concept mapping   name   Denny

water
solid
liquid
gas
vapor
river
ice
steam

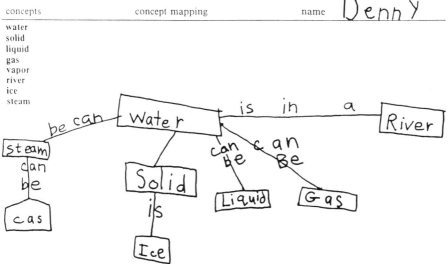

Figure 3.2. Concept Map Showing How a Six-Year-Old Child Represents Meanings for Concepts Related to Water

produced by a grade 4 student after a field trip to a paper mill, words presented to the class regarding the field trip were strung together with little or no meaning, especially later in the word string. This child was the best oral reader in her class, but she had almost no comprehension of what she read. It is not uncommon for children to be high performers in routine classroom activities and yet derive little or no meaning from the instruction. They can perform very well on typical classroom tests or assignments, but gain little or no understanding of basic concepts and principles. Concept maps can serve as a diagnostic tool to identify students who are suffering from a pattern of rote-mode learning.

When students are asked to construct concept maps as a routine part of their instruction, they can move from patterns of rote learning to patterns of meaningful learning. Although the initial experience can be agonizing for many students, we have not found a single case in which students who genuinely attempt to produce hierarchical structured concept maps fail to achieve patterns of meaningful learning. This has been true with elementary school students as well as students at Cornell University (Donn, 1990). In my course Learning to Learn, the majority of students entered the course as predominately rote learners and, after several weeks of struggling with concept mapping and other learning tools, all were successful in altering their learning patterns. Typically, they described the frustration of their previous unproductive patterns of rote learning and their anger with the system that encouraged them to learn in this way. However, they also recognized that attaining higher grades on most course examinations requires little more than rote learning, a practice with which they had become skillful over 12 or 14 years of schooling. There is considerable emotional upheaval for students who accept the challenge and move their learning patterns from memorization to the construction of meaningful learning.

DESIGN OF INSTRUCTION

Early in our experience with concept maps, we found that they were useful tools for organizing a lecture or an entire curriculum. We used concept maps to design a curriculum for specialists concerned with the application of waste water on land as a sewage treatment alternative with support from the Environmental Protection Agency. In fact, we floundered for a year trying to write outlines for instructional units and to plan a coherent instructional program. Once we began to utilize concept maps to organize the information, both for the program as a whole and for subunits in the program, we rapidly proceeded to the preparation of 22 instructional modules, some of which were at an introductory level and some at a more advanced and detailed level. The end product was a publication that still is being used for instruction in this field (Loehr, Jewell, Novak, Clarkson, & Friedman, 1979).

In recent years, many individuals have found the use of concept mapping a powerful tool for organizing lectures, papers, chapters, or curricula. Many science textbooks now include concept mapping as a regular activity to help students summarize what they have learned in textbook chapters. While only a small minority of teachers regularly use concept maps to organize their lessons, it is clear that this device has high promise for the improvement of both instructional materials and the writing of papers reporting research or other activities. Concept maps are playing an important role in a major curriculum revision currently under way in Cornell's College of Veterinary Medicine (Edmondson & Novak, 1992). They are being used to analyze the conceptual structure of case studies (see Figure 3.3) to ascertain what conceptual coverage is achieved by this case-based instructional program.

In a seminar program with practicing teachers, I found that all of the participating teachers, each of whom had years of experience in science teaching, expressed frustration with the organization of information for teaching. By learning to utilize concept maps to organize knowledge,

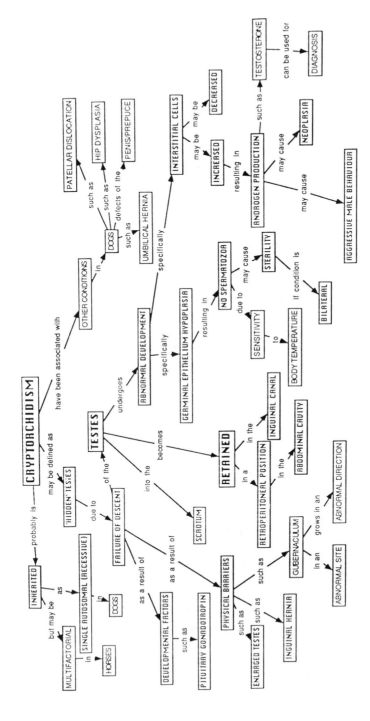

Figure 3.3. Concept Map from Case Study of Cryptorchidism (Undescended Testicles)

they found not only that they were aided in planning instruction, but also that their own understanding of the subject matter was increased. This was especially true in dealing with subject matter in areas less familiar to the teachers or in looking for relationships between different fields of science. The results of this seminar program were reported elsewhere (Novak, 1991). It should be noted, however, that the pressure of routine activities in the classroom, constraints of external examinations, and other factors have limited the extent to which these teachers have found it possible to apply concept mapping in their instructional programs and in work with students. Many factors in this school milieu conspire against practices that would encourage meaningful learning. Perhaps the over-riding factor limiting teaching effectiveness is the pressure of time. Teachers lack adequate time to prepare lessons, to design evaluation materials, to plan laboratory work, and to evaluate student performance. While concept mapping is no panacea, it can facilitate planning and evaluation and thus encourage practices that enhance meaningful learning.

Concept mapping has been utilized to design textbooks, including recent editions of the Biological Sciences Curriculum Study (BSCS) high school textbooks. It was found that concept maps served as a way to facilitate writing individual chapters, which were done by different authors, and achieving a conceptual integration across the whole textbook. These books also offer instructions to teachers for utilizing concept mapping in class instruction. A number of publishers now incorporate concept maps as part of the text, teacher's guide, or study guide.

DEALING WITH ALTERNATIVE CONCEPTIONS

Concept maps are a powerful tool for identifying students' misconceptions and also for facilitating learning to change conceptual understandings (Feldsine, 1983). Misconceptions or alternative conceptions are not isolated pieces of knowledge but rather a segment of a complex propositional framework. One of the reasons misconceptions are difficult to modify is that it is not clear which elements in an individual student's conceptual framework require alteration or additional concepts and propositions. I have suggested that we label misconceptions (or alternative conceptions) as *limited or inappropriate propositional hierarchies* (LIPHs) in recognition that we are dealing with complex segments of cognitive structure and not individual statements or definitions of concepts or principles (Novak, 1983).

In a recent study with university students in chemistry, we found that the use of concept maps revealed misconceptions or incomplete understanding even for students who scored well on written examinations (Pendley, Bretz, & Novak, 1994). For example, in one phase of our study with above-average undergraduate majors in chemical engineering, four out of six students who scored full marks on a written examination involving problem solving showed important misconceptions in interviews following the examination. All six students lacked an understanding of some of the basic principles involved in gas chromatography. There are useful ways in which concept maps can be employed to help students to recognize and overcome misconceptions, especially when they are used in small-group settings. In addition, concept maps are helpful to teachers, who also need to identify their own misconceptions and find pathways to correct their understandings. This is especially true at the elementary school level. For a useful summary of research on alternative conceptions in science, see Wandersee, Mintzes, & Novak (1994).

STUDENT EVALUATION

One of the most troublesome areas for teachers who hope to move students from memorization of information to understanding of science concepts and principles is the limitations of traditional evaluation tools. Multiple-choice tests are notoriously poor at measuring high levels of conceptual understanding, even though theoretically one should be able to test for evaluation and synthesis of ideas. In practice, most multiple-choice tests or other objective tests require little more than verbatim recall of specific information. To design a single multiple-choice test item that probes for understanding unambiguously can take an hour or more of a teacher's time. Because time is the perennial enemy for teachers, this is not possible, and examinations provided by publishers rarely achieve this level of evaluation (Holden, 1992).

Concept maps, on the other hand, can be utilized as powerful evaluation tools requiring high levels of synthesis and evaluation by simply asking students to map out a set of related concepts for any topic of instruction. If students are provided with 10 to 20 concepts to map for a given topic of study, they must evaluate which concepts are the most significant superordinate concepts and also determine the subordinate concepts and appropriate linking words to describe the concept relationships. This requires successive efforts at synthesis and evaluation, as well as knowledge of the specific concepts and their definitions. If students also are asked to add to their maps several more concepts that are related to the concepts given, the challenge of recall, synthesis, and evaluation is strengthened further. This form of evaluation serves also as a powerful teaching vehicle. It is relatively easy for a teacher to identify the 10–30 concepts that are relevant to any given topic.

Furthermore, student maps help the teacher to see misconceptions held by the student (Fraser & Edwards, 1983) and to strengthen the teacher's own conceptual understanding of the discipline. Of course, students need practice and experience in becoming skillful in concept mapping, and this requires patience on the part of both teachers and students. Scoring or evaluating concept maps can take up to 10 minutes per student, but this leads to other useful consequences. Concept maps also provide highly specific information to teachers in identifying specific places where the instructional program has failed to teach important concepts or propositions.

COOPERATIVE LEARNING

There is a growing recognition that learning in groups can be highly effective both as a teaching strategy and as a learning strategy. Instruction utilizing groups to organize, analyze, and solve problems, while still occurring only in a minority of classrooms, is proving to be a promising avenue for improvement of school instruction (Sharan, 1980; Slavin, 1987). One of the difficulties many teachers have in cooperative learning is how to focus students' attention on key issues to be understood and how to involve all members of the group in participation. Concept mapping can be effective by involving students in preparing individual maps of their understanding of the problem or case they are studying and then having them work collectively to merge their maps into a more comprehensive group map. This approach provides an agenda for lively dialogue and debate that is on task and leads to productive work. While there is little research on the value of concept mapping in cooperative learning programs, teachers employing this strategy report enthusiasm for this method of instruction.

COMMITMENT TO CONSTRUCTIVIST THINKING

It now is accepted widely that all knowledge is constructed by human beings and that knowledge is modified continuously as new ways are developed to observe and organize experience. One important educational goal should be to help students and teachers to understand the constructed nature of knowledge and the constantly evolving characteristic of meanings for explanatory models that we construct. Partly because concept maps show the infinite permutations of concepts and propositions that can be organized to explain any given phenomenon, they can be a powerful tool for helping students to understand the meaning of the constructed nature of knowledge (Novak, 1993). Of course, this is an important understanding that should be developed by teachers, many of whom suffer from the influence of highly positivistic approaches to teaching in college science, which tend to present science as a mass of facts to be memorized (Linn, Songer, & Lewis, 1991). Constructivist thinking not only facilitates meaningful learning, but also moves toward the very important goal of public understanding of science and the recognition that scientific knowledge undergoes continuous evolution. This kind of understanding can help future citizens recognize the importance of basic research and the need for constant scrutiny of evidence in any domain of knowledge. This goal of science teaching can be as important as the goal of gaining information on the major phenomena of science.

EMPOWERMENT OF TEACHERS AND LEARNERS

A common characteristic of average laypersons is that they lack confidence in their understanding of most areas of science. When interviewed about almost any phenomenon dealing with the natural world, most people will admit quickly their fragile understanding or that they have "forgotten" the answers to our questions. Having been schooled in science courses that required largely memorization and regurgitation of information, most people have failed to form conceptual frameworks that are powerful in constructing explanations for scientific phenomena. Moreover, they lack an understanding of complex concepts and, therefore, fail to control the quality of their own meanings, even in those domains for which they have significant factual information. This kind of disempowerment leads to apathy about science at best, and scepticism or even cynicism at worst. It is not only the lost effort of years of science teaching focused on memorization that is of consequence; it is also the very serious disempowerment that occurs from this kind of instruction that needs to be considered. Concept mapping can be a powerful tool to facilitate meaning-making and to facilitate a sense of personal control over meaning-making for future citizens.

CONCLUSION

Concept mapping has been found to be useful for improving science teaching and learning by facilitating meaning-making and providing a sense of personal control over this activity. Research reported in this chapter shows that concept maps can be a powerful tool for designing, organizing, and revising lectures, papers, chapters, or curricula. Further, concept maps can be used to identify students' misconceptions or alternative frameworks, to facilitate learning to challenge and change conceptual understandings, and to act as classroom evaluation tools that assess high levels of understanding. With growing attention in schools to effective teaching and learning, concept mapping has been found to facilitate cooperative learning and other kinds of group work.

REFERENCES

Donn, J. S. (1990). *The relationship between student learning approach and student understanding and use of Gowin's Vee in a college level biology course following tutorial instruction.* Unpublished doctoral dissertation, Cornell University, Ithaca, New York.

Edmondson, K. M., & Novak, J. D. (1992). Toward an authentic understanding of subject matter. In S. Hills (Ed.), *The history and philosophy of science in science education* (Vol. 1, pp. 253–263). Kingston, Canada: Faculty of Education, Queen's University.

Feldsine, J. (1983). Concept mapping: A method for detection of possible student misconceptions. In H. Helm & J. D. Novak (Eds.), *Proceedings of the International Seminar on Misconceptions in Science and Mathematics* (pp. 467–476). Ithaca, NY: Department of Education, Cornell University.

Fraser, K., & Edwards, J. (1983). Concept maps as reflectors of conceptual understanding. *Research in Science Education, 13,* 19–26.

Holden, C. (1992). Study flunks science and maths tests. *Science, 258,* 541.

Linn, M. C., Songer, N. B., & Lewis, E. L. (Eds.). (1991). Students' models and epistemologies [Whole issue]. *Journal of Research in Science Teaching, 28,* 729–882.

Loehr, R., Jewell, W. J., Novak, J. D., Clarkson, W. W., & Friedman, G. S. (1979). *Land application of wastes* (Vols. I & II). New York: Van Nostrand Reinhold Co.

Novak, J. D. (1983). Overview. In H. Helm & J. D. Novak (Eds.), *Proceedings of the International Seminar on Misconceptions in Science and Mathematics* (pp. 1–4). Ithaca, NY: Department of Education, Cornell University.

Novak, J. D. (1990a). Concept maps and Vee diagrams: Two metacognitive tools for science and mathematics education. *Instructional Science, 19,* 29–52.

Novak, J. D. (1990b). Concept mapping: A useful tool for science education. *Journal of Research in Science Teaching, 27,* 937–949.

Novak, J. D. (1991). Clarify with concept maps. *The Science Teacher, 58*(7), 45–49.

Novak, J. D. (1993). Human constructivism: A unification of psychological and epistemological phenomena in meaning making. *International Journal of Personal Construct Psychology, 6,* 167–193.

Novak, J. D., & Gowin, D. B. (1984). *Learning how to learn.* New York: Cambridge University Press.

Novak, J. D., & Musonda, D. (1991). A twelve-year longitudinal study of science concept learning. *American Educational Research Journal, 28,* 117–153.

Novak, J. D., & Wandersee, J. (Guest Eds.). (1990). *Concept mapping* [Special issue]. *Journal of Research in Science Teaching, 27,* 923–1075. (whole issue)

Pendley, B. D., Bretz, R. L., & Novak, J. D. (1994). Concept maps as a tool to assess instruction in chemistry. *Journal of Chemical Education, 71,* 9–15.

Sharan, R. E. (1980). Cooperative learning. *Review of Educational Research, 50,* 315–342.

Slavin, R. E. (1987). *Cooperative learning: Student teams* (2nd ed.). Washington, DC: National Educational Association.

Wandersee, J. H., Mintzes, J. J., & Novak, J. D. (1994). Learning alternative conceptions. In D. L. Gabel (Ed.), *Handbook of research on science teaching and learning* (2nd ed.) (pp. 177–210). New York and Toronto: Macmillan.

CHAPTER 4

Interviews About Instances and Interviews About Events

Malcolm Carr
University of Waikato, Hamilton, New Zealand

This chapter describes investigations of student understanding of science and mathematics through the use of interviews focused on instances (Osborne & Gilbert, 1979) or events (Osborne, 1980). Recent surveys of the use of these techniques are found in Carr (1991), White and Gunstone (1992), and Pfundt and Duit (1994). This method of exploring understanding and revealing the current concepts of students can be traced back to the clinical interviews developed by Piaget. The conversation with the interviewee typically centers around phrases such as "in your own understanding" and "in the way that you think about this," and the task is exploration of ideas rather than closed definitions. The interview-about-instances technique is based on discussions of cards containing pictures of familiar objects or occurrences in the world of the student. This conversation explores the student's concept by first focusing on whether the card portrays an instance or a noninstance of the concept being explored. The interview-about-events technique also stimulates conversation focusing on a concept, but this time the stimulus is an activity carried out with the student. A very helpful source of advice on many aspects of interviewing children is Bell, Osborne, and Tasker (1985).

This chapter provides a rationale and examples for the use of these techniques involving interviews about instances and events. The choice of appropriate interview foci is considered, hints about effective interviewing are provided, and the problems associated with interview procedures are discussed. The chapter ends with comments about the value of the conversation resulting from interviews.

A RATIONALE FOR THE USE OF INTERVIEWS

Many researchers in recent years have discussed the importance of the ideas that learners bring to lessons (Bell, 1993; Northfield & Symington, 1991; Osborne & Freyberg, 1985). Consequently, exploration of these prior ideas has become an important research focus. The task facing researchers of children's ideas is to describe as closely as possible the reality of the learner's ideas. This is not an easy task because the method used to explore these ideas can greatly influence responses by students (Jung, 1987; McLelland, 1984). The problem arises in two ways. First, because the options are chosen by the researcher rather than the student, the student can be provided with options that do not include a pathway to the concept that he or she holds. Second, the student's responses can be based on what the learner believes to be expected, rather than being genuinely a description of his or her ideas.

Because learners' ideas are constructed from a wide variety of prior experiences, it is important that the means of exploring them is capable of revealing this rich variety. This requires movement away from the frequently narrow context of school science and mathematics lessons. A typical feature of the interview about instances and the interview about events is that the exemplars explored range over familiar objects and occurrences in the world of the student. Exploration of the concept of force could depict a stalled car being pushed, and a warm saucer leaning on a cup to dry might be used to consider the concept of states of matter. An alternative probe might be a completely open conversation in which the researcher asks the student to describe the concept. However, such an approach is difficult to sustain, and often can lead to long and increasingly embarrassed silences. The unstructured interview is also confrontational for the interviewee. Direct eye contact and the alarmingly open nature of the interaction can cause conversation to flag, with the person being interviewed feeling that there is no place to hide.

The important contribution of Osborne and Gilbert (1979) to the exploration of children's science was to suggest a shared external focus of pictures or events that provide a comfortable focus for conversation ranging widely over the rich prior experiences of the learner. Subsequently, many concepts represented by pictures have been developed.

Interview procedures are valuable not only to researchers. Many teachers who have tried some of the classic interviews about concepts first developed by Osborne and Gilbert (1980) report that they have gained new insights into the ways in which their students think. Commonly, teachers report that they were unaware that their students already knew so much about a concept believed by the teacher to be new. Another frequent comment is that teachers were surprised at the amount of linkage to everyday ideas and events that students bring to the classroom. The remainder of this chapter suggests ways in which teachers and researchers can explore students' ideas by using these techniques.

THE INTERVIEW ABOUT INSTANCES

The interview-about-instances procedure uses a set of stimuli, most often cards with pictures and a small amount of writing. The interview begins with a request for the interviewee to respond with his or her own understanding of a focus concept by discussing the cards as they are revealed successively. As each card is shown, the interviewer asks such a question as: "In your own understanding of *the focus concept*, is this an example or not?" Follow-up questions then seek reasons for the initial response. The interviewer emphasizes throughout that there are no right answers, and the learner's ideas always are acknowledged and valued.

For example, through the interview-about-instances technique, the student's concept of "animal" can be explored using a series of cards with labeled pictures or line drawings of a person, a cow, a cobra, an eel, a monarch butterfly, an eagle, a mushroom, a flea, a robot, an elephant, a mouse, and a famous athlete. Similarly, the concept of "floating and sinking" could be explored through cards depicting a number of instances (which are labeled clearly) as illustrated in Figure 4.1.

THE INTERVIEW ABOUT EVENTS

The procedure for the interview about events is very similar to that of the interview about instances, with the stimulus being some activity carried out with the student. Again, the student's responses to questions about the concept being focused on is sought, and follow-up questions explore the reasons for the initial responses. Responses are acknowledged and valued. As an example of an interview about events, the concept of "states of matter" could be explored through the following activities: Some ice is melted in a saucer, a container of cold liquid becomes wet on the outside, a warm wet saucer is leaned against a cup to dry, an electric jug is boiled, a vitamin C tablet is dissolved in water, some butter is melted, some salt is dissolved in water, and a sealed plastic bottle full of warm air is cooled.

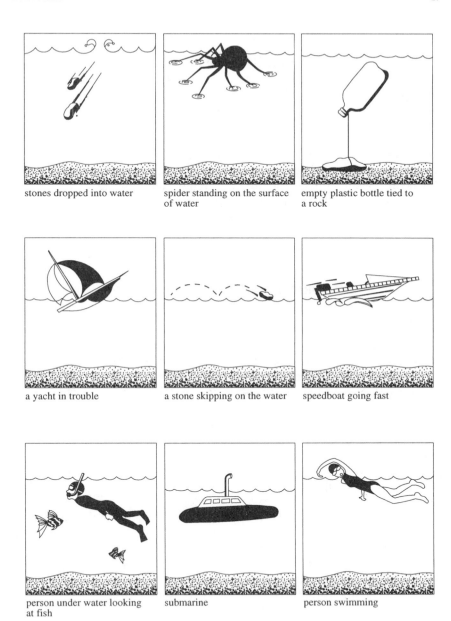

Figure 4.1. Cards Showing a Number of Instances of the Concept of Floating and Sinking

THE CHOICE OF INSTANCES AND EVENTS

The focus pictures or events in the examples above seek to engage students about their views of the science concepts. To explore these, it is necessary to move beyond the narrow focus of many science definitions and the typical examples used in school. This is exemplified by considering focus pictures to explore ideas of "living." Obvious choices are some animals and plants, but a greater richness of interaction with students' ideas comes from using the sun and a fire as further examples. The picture of a fire brought from many students the metaphoric use of this word. They said that a fire was living, and described how a fire "breathed" and "fed." When discussing changes of state, reference to a kettle boiling is preferable to discussion of water in a beaker heated by a bunsen burner. The condensation on the outside of a cold container of soft drink further links the interview with the real world of experience.

PROCEDURAL SUGGESTIONS FOR INTERVIEWS

Getting Started

The first decision has to be the choice of the focus for an interview. Most researchers and teachers are well aware of some concepts that are important to student understanding and that, experience suggests, cause difficulty. Published bibliographies (Carr, 1991; Pfundt & Duit, 1994) provide sources of concepts that previously have been explored to assist in developing the focus cards or activities for the interview. It often is valuable to try out the interview with a small group of students to deal with any ambiguities and other difficulties. This sometimes indicates that a picture or drawing is confusing because it contains unnecessary detail, and provides an opportunity to rehearse responses to comments that have not been expected.

In the interviews, some time should be spent helping the student to relax. Starting with conversation about a neutral topic, sitting alongside the student, and openly explaining why the interview is being held are all methods of beginning in an easy way. It can be helpful to ask questions that the student can answer without difficulty. When a teacher is interviewing students from his or her classroom, there is a particular need to be open. An example of a way in which to start an interview is given below:

> I suppose you are wondering what this is all about. I'm hoping that you can help me to understand the kind of thinking about ideas in

science that you bring to classes. This is because I want to think about how to be a better teacher by reacting to students' ideas. In all of our talk, try to forget that I am a teacher. As strange as it might seem, I am not looking for answers that I already know. This is not some kind of a test. I really want to hear your ideas in your own words. There are no right answers. Whatever you tell me will be very helpful as I try to think about teaching.

Nondirection of Responses

The aim of these interviews is to have the student reveal his or her thinking about a concept. This is done by reinforcing the need for the students to respond with their own ideas, and by emphasizing that the questions are not a test of knowledge. Questions such as "In your own meaning of the word, is this an animal?" and follow-up questions such as "Can you explain to me why *you think* that is an animal?" keep the emphasis focused on students' ideas. Sometimes an unexpected answer can surprise the interviewer. A useful response is to reflect it back by repeating it as the interviewer recovers from the surprise:

INTERVIEWER: Does the rainbow make light?
STUDENT: People say God made it.
INTERVIEWER: (*After slight pause*) People say God made it?. . . Can you tell me more about that?

Remember that questions can be closed or open. Closed questions keep you in control by implying the answer ("In your meaning of the word, is a fire living?") and will not reveal students' ideas as effectively as open ones ("Why do you say that?"). However, because open questions are more threatening than closed questions to the person interviewed, a balance needs to be struck between the two types. It is helpful to mix your interview with simple and direct questioning, and to judge when the student is becoming distressed or intimidated.

Encouragement

Students often take some time to believe that their responses are not being judged as right or wrong by the teacher. Of particular importance is the establishment of the rules of the game being played in an *interview.* Students are familiar with the rules of the *test* in which answers are right or wrong. Students can assume that an interview seeks the right answer from the teacher's or scientist's point of view. During the interview, phrases such as "in the way that you think about this" reinforce the idea that the interviewer is not seeking right answers. The interviewer can affirm the value of responses by comments such as "Thank you, that is helpful," and by nonverbal signals such as nodding and smiling. The great difficulty for many interviewers is to put aside their teaching strategies and accept curious responses without flinching or correction. It is vital to stay neutral even under provocation. A further related point is that the student should feel free to question the meaning of a question during the interview. This can be triggered by asking whether the question makes sense to the student.

Patience in Awaiting a Response

Some interviews take time to become fruitful. It does not help to rush, or to keep talking during what might be valuable silences when the student is formulating a reply. Helpful comments to break a silence could be "Remember I am interested in your own ideas here, so take your time," and "I'll put that question in another way." Rushing on is also unhelpful after early responses that are often the least revealing of a student's ideas, particularly when they are negative or an indication of uncertainty. The student could have returned to thinking that a right answer is sought. It might help to say "Can you tell me why you said no?" or "What are you not sure about?"

Cross-Referencing During the Interview

Effective interviews require an alert response from the interviewer. Careful listening can often provide clarification of students' ideas by reference to other aspects of the interview. At all times, avoid putting words into the learner's mouth and refer to the previous idea as accurately as possible. An example would be: "In your response concerning the monarch butterfly you said that it was not an animal because it was an insect. You just have said that a flea is an animal. Could you tell me a bit more about your thinking here?"

Recording and Interpreting the Information

It is impossible to remember all of the complex ideas explored during a lively interview, and attempting a written record takes attention away from careful listening and responding. The solution is to tape-record the interaction. This is best negotiated by saying that you want to be able to think about the conversation, explaining the fallibility of your recall, and asking to tape record it. The tape recorder is turned on without making a fuss as soon as the need for recording has been explained, and permission given. The tape recorder should be placed in an unobtrusive position, preferably near the interviewer and reasonably out of sight, and ignored thereafter.

The next problem is how to extract information from the record. Ideally the entire interview should be transcribed. Researchers find that this brings out the richness of the interaction and often reveals material that listening does not uncover. Busy teachers, however, are unlikely to have time for detailed transcribing of all tapes. The solution is to listen carefully to each tape several times, noting sections that warrant transcribing in full, and making notes about other less revealing sections. It pays to process the tape as soon as possible after the interview because memory fades rapidly.

There is an extensive literature on the interpretation of interview data, and this acknowledges the subjectivities that can affect this process (see, for example, Cohen & Manion, 1989; and Chapter 2, this volume). The traditional research concepts of reliability and validity need reconsideration in this qualitative research methodology. As is indicated in Chapter 2, the best process of analysis might be a combination of letting the data speak for themselves, and using the data to test an already-articulated hypothesis. Teachers and researchers who explore their students' understandings through interviews need to be aware of these complexities. Nevertheless, teachers who use interviews often state that the directness of the responses they obtain is valuable and illuminating for their teaching.

Problems in Interviewing Students

Although the interviews described here are reasonably nonthreatening, it is important to recognize that some students could be intimidated by the most sensitive approach, or they might be reluctant to answer questions from a mistaken belief that they are on trial. Reactions to exploring and challenging ideas can be different for different cultures, and shy students could become increasingly uncomfortable as the interview goes on. The best solution for an interview that gets stuck is to escape gracefully, with thanks for the time that the student has spent, and without recrimination.

CONCLUSION

The interviews about instances and interviews about events described in this chapter are valuable for helping teachers gain insight into the ideas of individual students. The value comes from entering into a conversation and listening very carefully to the replies. Procedures for doing this in an effective manner have been described in this chapter. The technique of setting up a conversation and listening to the resulting ideas can be transferred to the classroom in a number of ways. Individual items from an interview schedule could be used with a whole class to begin a brainstorming session. Groups of students could be encouraged to discuss some of the instances or events, and then to report their ideas to the whole class. Other techniques for investigating students' concepts are explored in Chapters 3, 5, and 6 in this volume. In addition to gaining insight into the concepts held by students, teachers should be alert to the strength with which these concepts are held; some concepts are embedded strongly in students' thinking about their world (as in the case of electric current in circuits, described by Cosgrove, Osborne, & Carr, 1985) and others are not strongly held (as in the case of photosynthesis, described by Barker & Carr, 1989). Teaching approaches that interact with existing concepts can vary depending on the degree of commitment to these concepts shown by students.

Alternative frameworks for concepts in science and mathematics provide important information about students' existing knowledge. Listening to and acknowledging these ideas is important. Even more important is building teaching from them. A number of effective teaching and learning approaches have been developed to work from existing concepts by generating new ideas from them, and by challenging existing ideas that conflict with the concepts of science (Biddulph & Osborne, 1984; Osborne & Freyberg, 1985; Osborne & Wittrock, 1985).

REFERENCES

Barker, M., & Carr, M. (1989). Teaching and learning about photosynthesis: Part I. An assessment in terms of students' prior knowledge. *International Journal of Science Education, 11*, 49–56.

Bell, B. (1993). *Children's science, constructivism and learning in science*. Geelong, Australia: Deakin University Press.

Bell, B., Osborne, R., & Tasker, R. (1985). Finding out what children think: Appendix A. In R. Osborne & P. Freyberg (Eds.), *Learning in science: The implications of children's science* (151–165). Auckland, New Zealand: Heinemann.

Biddulph, F., & Osborne, R. (1984). *Making sense of our world: An interactive teaching approach.* Hamilton, New Zealand: University of Waikato.

Carr, M. D. (1991). Methods for studying personal construction. In J. Northfield & D. Symington (Eds.), *Learning in science viewed as a personal construction: An Australasian perspective* (pp. 16–24). Perth, Australia: Curtin University of Technology.

Cohen, L., & Manion, L. (1989). *Research methods in education* (3rd ed.). London, England: Routledge.

Cosgrove, M., Osborne, R., & Carr, M. (1985). Children's intuitive ideas on electric current and the modification of these ideas. In R. Duit, W. Jung, & C. von Rhoneck (Eds.), *Aspects of understanding electricity: Proceedings of an international workshop in Ludwigsburg, 1984* (pp. 247–256). Kiel, Germany: Schmidt & Klaunig.

Jung, W. (1987). Understanding students' understandings: The case of elementary optics. In J. Novak (Ed.), *Proceedings of the Second International Seminar on Misconceptions and Educational Strategies in Science and Mathematics* (Vol. III, pp. 268–277). Ithaca, NY: Cornell University.

McLelland, J. (1984). Alternative frameworks: Interpretation of evidence. *European Journal of Science Education, 6,* 1–6.

Northfield, J., & Symington, D. (Eds.). (1991). *Learning in science viewed as a personal construction: An Australasian perspective.* Perth, Australia: Curtin University of Technology.

Osborne, R. J. (1980). Some aspects of the students' view of the world. *Research in Science Education, 10,* 11–18.

Osborne, R., & Freyberg, P. (1985). *Learning in science: The implications of children's science.* Auckland, New Zealand: Heinemann.

Osborne, R. J., & Gilbert, J. K. (1979). Investigating student understanding of basic physics concepts using an interview-about-instances approach. *Research in Science Education, 16,* 40–48.

Osborne, R. J., & Gilbert, J. K. (1980). A method for investigating concept understanding in science. *European Journal of Science Education, 2,* 311–321.

Osborne, R., & Wittrock, M. (1985). The generative learning model and its implications for learning science. *Studies in Science Education, 12,* 59–87.

Pfundt, H., & Duit, R. (1994). *Bibliography—Students' alternative frameworks and science education* (4th ed.). Kiel, Germany: Institute for Science Education (IPN), University of Kiel.

White, R., & Gunstone, R. (1992). *Probing understanding.* London, England: Falmer Press.

CHAPTER 5

Computer-Video–Based Tasks for Assessing Understanding and Facilitating Learning in Geometrical Optics

Fred Goldberg and Sharon Bendall
San Diego State University, California, United States

Previous research on student learning in geometrical optics has indicated that students exhibit difficulty when asked to provide qualitative explanations and draw ray diagrams to account for various phenomena involving light (Bendall, Galili, & Goldberg, 1993; Galili, Bendall, & Goldberg, 1993; Goldberg & McDermott, 1986, 1987). Many of these difficulties emerge when students are asked to make explicit connections between the basic principles of geometrical optics and specific phenomena they observe. In particular, students have difficulty taking into account the observer in both their verbal explanations and their diagrams. (For example, a student might be able to locate the position of an image in a plane mirror, but might have difficulty determining how much of the image would be seen by an observer at a given position.)

Over the past several years, the Physics Learning Research Group at the San Diego State University Center for Research in Mathematics and Science Education has attempted to address the difficulties mentioned above by developing an instructional strategy that utilizes computer-videodisc systems. We have used our approach mainly with college students enrolled in activity-based physics courses for prospective elementary teachers, and in situations in which the instructional focus is on helping the students develop a good conceptual understanding of geometrical optics. The strategy is built around both a set of demonstration tasks administered by the computer-videodisc system and the capability of the system to overlay graphics and text on top of video pictures. In this chapter, discussion focuses first on some of the characteristics of the successful demonstration tasks and then on their use with the computer-videodisc system in assessing students' knowledge and facilitating students' learning.

TASK DEVELOPMENT

Good tasks are simple and unambiguous, and have outcomes that are often surprising to students. The tasks we have found useful involve a set of questions about simple optical systems (lenses, mirrors, shadows, etc.). We ask students for predictions about what would happen if the optical system were to change in some prescribed way, and then ask for explanations of how students are thinking. The tasks have been administered both to pre-instruction students to elicit their prior knowledge and to postinstruction students to bring out their conceptual difficulties. Below are two examples of our tasks, one involving a plane mirror and the other involving a real image formed by a converging lens.

The plane mirror task is aimed at eliciting the student's ideas about how a plane mirror works. The student stands about one meter away from a small vertically mounted (or vertically held) square plane mirror, about 30 centimeters on each side, and is asked how much of his or her body he or she can see currently in the mirror. After a response (e.g., that he or she can see from forehead to waist), the following question is posed: "Is there anything that you could do to enable you to see more of yourself in the mirror?" The vast majority of students who are asked this question respond that they can see more of themselves by moving further back, and most claim to remember having done it before. After they explain why they think that the mirror works that way, we usually let them step back and observe that, much to their surprise, moving away does *not* enable them to see more of themselves (Goldberg & McDermott, 1986).

The converging lens task (Goldberg & Bendall, 1992; Goldberg & Mc-Dermott, 1987) elicits pre-instruction students' ideas about how a lens works and postinstruction students' understanding of the special light rays used in constructing a ray diagram. A self-luminous object, a converging lens, and a screen are set up so that there is an inverted image of the object on the screen. The learner is asked: "Suppose that the upper half of the lens is covered with a card. What would be seen on the screen?" (See Figure 5.1.) Both pre- and postinstruction students often respond that only part of the image would be seen on the screen because part of the image would be blocked by the card. The pre-instruction students seem to think of the image-formation process in terms of an entire

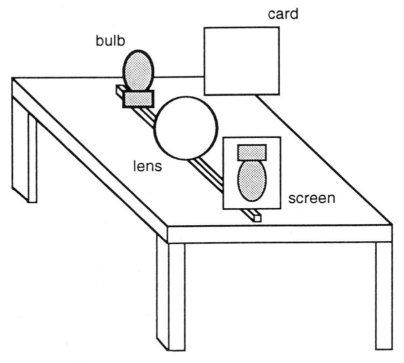

Figure 5.1. Apparatus Setup Showing the Real, Inverted Image of the Bulb on the Screen

image going through the lens and to the screen and that, with the card in place, part of that "potential image" would be blocked (Goldberg & McDermott, 1987). The postinstruction students justify their belief by pointing out that one or more of the (special) light rays that would go from the object to the image would be blocked, and hence part of the image would not be formed (Goldberg & McDermott, 1987). Throughout this demonstration, however, the entire image remains on the screen, even when the card is covering part of the lens. The image just gets dimmer as the card prevents some of the light from the object from contributing to it.

Over the years, we have developed a large number of tasks similar to those described above. Generally, the tasks first are developed and refined in clinical settings with individual students. Then, a set of tasks is videotaped and a videodisc for use with a computer-videodisc system is produced. In the next section, the system is described.

COMPUTER-VIDEODISC SYSTEM

Our computer-videodisc systems use a graphics-on-top-of-video feature that allows graphic ray diagrams to be drawn while superimposed on video stills or scenes. The choice to utilize graphics-on-top-of-video was based on psychological research (Tulving, 1983; Tulving & Thompson, 1973) suggesting that linkages between different types of knowledge can be strengthened if retrieval cues are established at the time the new information is encoded in memory. This implies that, for students to become adept at connecting ray diagrams with optical phenomena, they should learn the rules of ray diagram construction while they are viewing the optical phenomenon. With our system, the students can encode both images of the ray diagram and the actual phenomenon at the same time.

Our use of the computer-videodisc system has passed through two phases. In the first phase, we used the system to present tasks to students solely to assess their knowledge. In the second phase, we incorporated the tasks into a series of learning activities. Each of these two phases is discussed below.

Using the Computer-Videodisc System to Assess Student Knowledge

We developed a program that uses the computer-videodisc system to assess individually students' understanding of certain optical phenomena. Both pre- and postinstruction students were involved. Tasks were presented via the videodisc and the responses were analyzed by computer. For example, in the converging lens assessment, the narrator introduced some apparatus and proceeded through a series of questions inquiring about how the image might change if the setup were changed in particular ways. As each question was posed, the task was demonstrated in its entirety to eliminate any ambiguity in the question, but the actual outcome was hidden from the viewer's eyes. For instance, in the converging lens task described earlier in the chapter (also see Figure 5.1), as the narrator lowered a card over the lens, the translucent screen in the optical setup was covered with computer-generated graphics to block out what actually would be seen. After each question was posed, the students were directed to sketch on paper what they predicted would appear on the screen, and to draw on paper a ray diagram to justify the prediction. Finally, a video montage of common student predictions, as identified by previous research (Goldberg & McDermott, 1987), was shown. Figure 5.2 shows a graphic representation of this montage. Each of the possible screen pictures in the montage, except the one corresponding to the *actual* outcome of the task, was photographed under *contrived* conditions to produce the effect shown in the montage.

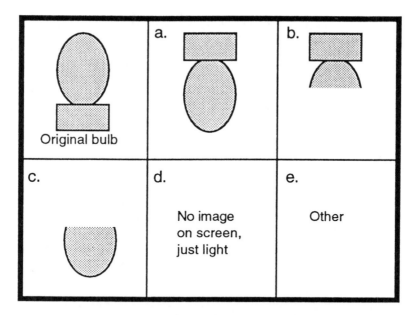

Figure 5.2. Example of a Montage of Common Student Responses

The student was asked to choose from the montage the screen picture that most closely corresponded with his or her prediction. Choosing from the montage provided an avenue for students to communicate with the computer system. The converging lens assessment contained a sequence of five to nine tasks. The exact number of tasks depended on the student's responses to the first few questions. After the final task, each student was given the option of observing scenes from the videodisc that showed the outcome of each task. The entire session lasted between 30 and 45 minutes.

A major strength of the computer-videodisc system was that it allowed us to ask students questions without a facilitator being present. However, we originally were concerned that this assessment tool could influence student responses. To address this concern, we first compared the responses to several of the questions included in the computer interview with those from a different study in which the first author asked the same questions to students in individual interviews (Goldberg & McDermott, 1987). We found that the type and pattern of student responses were similar in both situations (Goldberg, 1987). Second, we thought that perhaps the use of the video montage might encourage students to change their minds, but this was not the case. Students rarely made a choice that was inconsistent with their written prediction, and the montage rarely caused them to change a written prediction. (On their response sheets to the computer interviews, students were asked not to erase answers, but to note changes in their predictions in the appropriate spaces provided.)

After assessing student knowledge, we began the process of using this information to develop instructional strategies to facilitate student understanding in geometrical optics. A key component of our current instruction is a set of computer-videodisc programs that build on the demonstration tasks used in the computer-videodisc assessment, and complement hands-on activities in a physics course for prospective elementary school teachers.

Using the Computer-Videodisc System to Facilitate Learning

To integrate the computer-videodisc system into our entire set of activities in geometrical optics, we needed to expand the range of tasks presented by the system. To accomplish this, we videotaped additional tasks and produced a new videodisc. The tasks focused on the following topical areas: illumination on a screen; simple and complex shadows; pinhole patterns of light; distortion of objects viewed through different transparent media (like the apparent bending of a stick partially immersed in water); and images formed by plane mirrors, concave and convex mirrors, and converging and diverging lenses. There was another major difference between the new and original videodiscs. On the new disc, most dynamic sequences and still pictures actually show two simultaneous views of each optical setup. (They had been videotaped with two cameras.) One view is an "eye view" of the phenomenon. For example, this might be a view of a screen with an image on it formed by a converging lens, or a view of what is seen in a plane mirror from a particular position. The second view is the "orthographic view," usually the top or side view of the apparatus. It is with this view that the ray diagram is constructed most conveniently. Figure 5.3 shows what this dual view looks like for a converging lens setup. In the sketch, the eye view (screen view) in the upper right was taken with a camera pointed directly at the screen.

In addition to overlaying graphics on video, our Macintosh computer hardware and software allow students to draw their own ray diagrams by using a mouse-controlled cursor. When pedagogically appropriate, it is also possible to change the size of the video images in order to show more than one video still at a time, or show video stills and a dynamic scene (all of which are stored at different locations on the videodisc).

The environment in which our physics course is taught is a two-room complex. One room serves as a wet laboratory where students perform

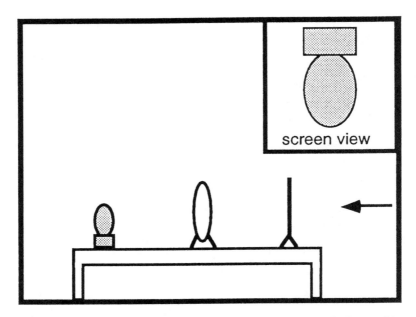

Figure 5.3. Dual Views of the Converging-Lens Setup: The Orthographic View (Side View) and the Eye View (Screen View)

experiments and engage in classroom discussions. A connecting room houses 14 computer-videodisc systems. The computer-video activities are designed to be used by students working in groups of two or three. Our course for prospective elementary school teachers is limited to an enrollment of 30 students, so that there are enough stations for each group to have a computer-videodisc system.

The computer-video activities that we have developed are built around the demonstration tasks. They are designed to guide students to an understanding of key ideas in geometrical optics, particularly the construction and interpretation of light ray diagrams. To facilitate this process, several strategies expected to promote learning have been incorporated.

LEARNING STRATEGIES INCORPORATED INTO THE COMPUTER-VIDEODISC LESSONS

The computer-video activities complement hands-on laboratories and class discussions. For instance, before working through the computer-video activity on real images formed by a converging lens, students first would observe an actual setup. Next, they would work through the computer-videodisc unit and, finally, they would return to the actual setup to reproduce some of the video-based tasks demonstrated during the activity and to carry out a set of additional experiments.

The tasks presented at the beginning of each activity are designed to elicit prior knowledge. At crucial points, students are asked to ponder certain issues, and to discuss their reasoning with their partners (see Figure 5.3). This process helps the students to make explicit their initial thinking about the phenomenon, a process that is suggested to be important for learning according to constructivist principles (Scott, Dyson, & Gater, 1987). We have found that the discussion surrounding each request to ponder usually lasts from one to several minutes. Additional tasks presented throughout the activity provide problem-solving contexts in which students develop a need to expand and apply their knowledge (Anderson, Boyle, Corbett, & Lewis, 1990). Furthermore, the outcomes of many of the tasks are surprising to students, and this provides a rationale for students to modify their thinking.

The computer-video activities incorporate "dual encoding" to address the difficulties students have in making explicit connections between optical phenomena and diagrammatic representations of those phenomena. On the computer monitor, the students see simultaneously an orthographic view of the apparatus and the eye view. The students then use a mouse-controlled cursor to construct the appropriate ray diagram directly on top of the orthographic view of the apparatus. (The feedback provided to the students also shows the expert diagram drawn on top of the orthographic view.) This process helps to make explicit the connection between the diagrammatic representation and the corresponding optical phenomenon. A representation of an observer's eye is sometimes included graphically on the orthographic view of the apparatus. In these cases, the student's diagram must show explicitly how light enters the observer's eye. In the case of virtual images, the student's diagram also must show where the virtual image is perceived to be located. This helps the student to make another crucial connection, namely, that between the diagrammatic representation of the observer and what the observer actually "sees." The latter view also is shown explicitly on the computer monitor.

In observing our students working through the activities, we find in general that they seem to be involved very actively. They often engage in animated discussions about the task questions, and they often construct "aerial diagrams" with their hands in front of the computer monitor before using the mouse-controlled cursor to draw the diagrams on the screen. Very rarely have we seen such gesturing when students are diagramming using pencil and paper or when they are working with actual laboratory apparatus. This gesturing indicates that the learners are quite involved in trying to understand a phenomenon and to communicate their thinking to their partners. In this case, the computer seems to act as a mediator to promote discussion between students.

The computer-video activities offer various kinds of feedback. Our system does not evaluate the students' diagrams. Instead, after constructing their own diagrams to account for some phenomenon, students usually are shown an expert diagram with key features circled and described. The choice of which features to highlight is based on our research on student difficulties in understanding these phenomena. In addition to diagrammatic feedback, we follow some questions with audio segments accompanied by simple animations.

Finally, at appropriate times during the activity, ideas and prototype diagrams are summarized on the computer screen. Students are directed to consider these as major ideas and diagrams. At later points in the computer-video activity, the students are asked to make use of these previously developed ideas to guide their thinking when responding to new tasks. We want students to recognize that new knowledge builds on previous knowledge and that they can develop a deep conceptual understanding of geometrical optics in terms of a limited number of powerful ideas and prototypical diagrams.

Assessing Student Understanding of the Principles of Geometrical Optics

During the past several years, we have conducted a number of studies to assess student understanding of the principles of geometrical optics. In each of these studies, we have carried out individual interviews with students enrolled in a physics class for prospective elementary teachers. The first study was carried out before we introduced the computer-videodisc programs into the curriculum (Galili et al., 1993), and the second was carried out after we introduced the initial version of the programs (Goldberg & Bach, 1991). We asked similar questions in both sets of interviews, as well as questions that were different. Each of the questions, however, was intended to be novel to the students. Furthermore, because of different research goals, we analyzed the data collected in the second study in a way different from that used in the data collected during the first study. Nevertheless, the two studies allow us to make some comparisons of the students' knowledge following the corresponding instructional units in geometrical optics.

In both studies, students were asked to observe phenomena involving actual apparatus and to draw ray diagrams to account for the phenomena. One task common to both studies involved the real image of a bulb formed with a concave mirror. The other common task involved seeing two virtual images of an object when looking at it through the apex of a triangular prism (Galili et al., 1993). In the first study, carried out prior to the inclusion of the computer-videodisc programs in the curriculum, only 8 out of 25 students interviewed (32 percent) included in their concave mirror diagrams a majority of the features present in an expert diagram. For their prism diagrams, only 3 out of 13 students (23 percent) included a majority of the essential features. In the study conducted following the introduction of the computer-videodisc materials, on average students' diagrams for the concave mirror task included 70 percent of the essential features present in the expert diagram, and on the prism task they included 86 percent of the essential features. Overall, then, students who had the computer-videodisc instruction drew diagrams that were more accurate than those drawn by students who did not have the opportunity to work with the computer-videodisc materials.

CONCLUSION

In our work, we have used carefully developed tasks to assess understanding and encourage learning. Key features of the tasks include simplicity, clarity, and a surprising outcome. Although the tasks originally were developed during our research on student understanding of geometrical optics, we now use them as the cornerstone of a series of computer-videodisc activities that incorporate numerous learning strategies. Specifically, the surprising outcome of the tasks motivates students to try to develop an understanding of the behavior of light in various circumstances. The development of this understanding is guided by a *series* of appropriately sequenced tasks. Each task promotes the students' development of key ideas. We have used the combination of video and graphics to help students to connect the actual phenomenon with the representation (i.e., ray diagrams) that helps to account for that phenomenon. Students work in small groups of two or three to take advantage of the benefits of social interaction in learning. The computer activities that we have developed complement other classroom activities to form a powerful learning environment in a physics course for prospective elementary teachers.

ACKNOWLEDGMENTS

The authors wish to acknowledge the work of various members of the Physics Learning Research Group at the San Diego State University Center for Research in Mathematics and Science Education. Those other members who have contributed to the optics project are Igal Galili, Sharon Streight, Niña Bach, Glenn Bach, Jim Meyer, and James Sammer.

REFERENCES

Anderson, J. R., Boyle, C. F., Corbett, A., & Lewis, M. (1990). Cognitive modelling and intelligent tutoring. *Artificial Intelligence, 42*, 7–49.

Bendall, S., Galili, I., & Goldberg, F. (1993). Prospective elementary teachers' prior knowledge about light. *Journal of Research in Science Teaching, 30*, 1169–1187.

Galili, I., Bendall, S., & Goldberg, F. (1993). The effects of prior knowledge and instruction on understanding image formation. *Journal of Research in Science Teaching, 30*, 271–301.

Goldberg, F. (1987). Using an interactive videodisc as a tool for investigating and facilitating student understanding in geometrical optics. In J. Novak (Ed.), *Proceedings of the Second International Seminar on Misconceptions and Educational Strategies in Science and Mathematics* (Vol. III, pp. 180–186). Ithaca, NY: Cornell University.

Goldberg, F., & Bach, N. (1991, April). *A strategy for assessing the extent to which students can recall and apply scientific knowledge.* Paper presented at the annual meeting of the National Association of Research in Science Teaching, Fontane, WI.

Goldberg, F., & Bendall, S. (1992). Computer-video-based tutorials in geometrical optics. In R. Duit, F. Goldberg, & H. Neidderer (Eds.), *Proceedings of the International Workshop on Research in Physics Learning: Theoretical issues and empirical studies* (pp. 356–379). Kiel, Germany: Institute for Science Education (IPN), University of Kiel.

Goldberg, F., & McDermott, L. C. (1986). Student difficulties in understanding image formation by a plane mirror. *The Physics Teacher, 24*, 472–480.

Goldberg, F., & McDermott, L. C. (1987). An investigation of student understanding of the real image formed by a converging lens or concave mirror. *American Journal of Physics, 55*, 108–119.

Scott, P., Dyson, T., & Gater, S. (1987). *A constructivist view of learning and teaching in science.* Leeds, England: Children's Learning in Science Project, Centre for Studies in Science and Mathematics Education, University of Leeds.

Tulving, E. (1983). *Elements of episodic memory.* New York: Oxford University Press.

Tulving, E., & Thompson, D. M. (1973). Encoding specificity and retrieval processes in episodic memory. *Psychological Review, 80*, 352–373.

CHAPTER 6

Using Teaching Experiments to Enhance Understanding of Students' Mathematics

Leslie P. Steffe and Beatriz S. D'Ambrosio
University of Georgia, Athens, United States

Many in the mathematics and science education community today agree that knowledge is not received passively but is actively built up by students (von Glasersfeld, 1989). For example, the Commission on Standards for School Mathematics (1989) of the National Council of Teachers of Mathematics in the United States stated that "this constructive, active view of the learning process must be reflected in the way much of mathematics is taught" (p. 10).

Active learning is a core constructivist principle. But the principle is not restricted to students. In a truly constructivist teaching-learning process, the teacher, too, is an active learner. The teacher is learning about the students' mathematical understanding and, in the process, reconstructing his or her own mathematical understanding. Self-reflexivity, another principle of constructivism, refers to the fact that constructivists apply the principles of constructivism first and foremost to themselves in their activities (Steier, 1995). Confrey (1990) exemplified active learning and self-reflexivity as she described her learning of the conceptions of exponential functions held by her students: "Through the process of the interview, my own conception of exponential functions was transformed, elucidated, and enriched" (p. 129). Confrey was learning actively the mathematics of students. Her statement is reason enough to reformulate the principle of active learning to include mathematics teachers: *Knowledge of the mathematics of students is not passively received, but is actively built up by teachers.* This reformulated principle is a basic assumption of the constructivist teaching experiment.

In this chapter, we describe a teaching experiment in which the goal was to understand students' construction of multiplication. We show how the teaching experiment served to provoke modifications of the researcher's understanding of a particular student's ways of knowing and operating, as he sought to build a model of the student's mathematics.

CONSTRUCTIVIST TEACHING EXPERIMENTS

We engage in constructivist teaching experiments to learn the mathematics of our students. To achieve this goal, our firsthand experience of bringing forth, sustaining, and modifying the mathematics of students is indispensable. It is our goal to learn from our students because we believe there is no one else who can teach us how they think mathematically.

We believe that students have mathematical knowledge distinct from our own. How students interpret the problems that we give them, the ways and means that they have available to attempt solutions, and how they modify those ways and means all play fundamental roles in our learning about their mathematical knowledge. Our methodology is compatible with Piaget's use of clinical interviews, but an important divergence exists in that we use teaching episodes rather than clinical interviews over extended time periods. Collectively, these teaching episodes constitute a teaching experiment (Cobb & Steffe, 1983; Hunting, 1983; Steffe, 1983).

A teaching experiment is not an investigation of how to teach a predetermined or accepted way of operating mathematically. Instead, it is an exploratory tool derived from Piaget's clinical interview and aimed at providing understanding of what might go on in children's heads as they engage in mathematical activity. Because it involves experimentation with the ways and means of influencing students' mathematical activity, the teaching experiment is more than a clinical interview. The clinical interview is aimed at establishing where students are, whereas the teaching experiment is directed toward understanding the progress that students make over an extended period of time. Sinclair (1988) pointed out that, methodologically, the teaching experiment is original as we use it: "It is a longitudinal but not a naturalistic study, since the experimenter-teachers [direct] their interaction with each individual child with a view to his or her possible progress" (p. v).

In the study of students' mathematical concepts, we regard our own mathematical concepts as starting points. But, as Confrey pointed out, we should not expect our concepts to be identical to those of the students. Rather, we expect to learn the students' concepts as modifications of our own. In a teaching experiment involving students' construction of multiplication, the experimenter-teacher started with a concept of multiplication quite similar to that described by Davydov (1991). According to Davydov, finding out how many metal coins are stored in a box usually involves weighing the coins rather than counting them, with the number of coins per unit weight being determined prior to weighing. Davydov sees the fundamental mental operation of multiplication in assembling coins into weight units: "The inconvenience . . . of a practical solution to this problem required a change of units—a transfer to larger units. This constitutes *the chief operation* of multiplication" (1991, p. 27).

However, there is no hint anywhere that his analysis of multiplication included students' thought. Teaching experiments into students' understanding of multiplication have led us to reconceive our view of multiplication to include the students' ways of thinking and operating. The reader is reminded that, throughout a teaching experiment, researchers draw inferences about the students' mathematics as they observe students solving problems and listen to students explain their solutions. The inferences drawn and explanations sought reflect the lens through which the experimenter observes and interacts with the students. At the end of the teaching experiment, the experimenter-teachers engage in a retrospective analysis of the teaching episodes. This analysis results in coherent explanations of the students' mathematical operations and schemes and of the students' modifications of these operations and schemes.

An important feature of teaching experiments rests in the activities and situations used for the purpose of understanding the students' mathematics. The investigators create a context that they believe will elicit actions from the students. Embedded in the contexts is the investigators' anticipation of students' ways of thinking and operating. During a teaching episode, the situations are posed to the students, who in turn act on the situations in search of a solution. These actions and subsequent discussions of the actions reflect the students' interpretations of the situations at hand, though in retrospective analysis of the teaching episodes, the design of the situations, the investigators' anticipation of students' actions, and the solutions offered by the students all come to bear on an evolving model of the students' mathematics.

In the following sections, we describe a teaching experiment conducted with Maya, an eight-year-old girl (Steffe, 1992). We describe situations that were posed to the child during the teaching episodes and discuss what was learned through retrospective analysis of those teaching episodes.

UNITS-COORDINATING SCHEMES

We claim that, for elementary school children, multiplication arises as a modification of their existing number sequences, which appear in children's mathematical activity as counting schemes. For example, Maya was asked to place eight marbles into an opaque cup and seven marbles into another. She then was asked how many marbles were in both cups together. After she counted and said "fifteen," she was asked to place three more marbles with the seven and then was asked how many marbles were now in that cup. She looked straight ahead before saying "ten," presumably counting "eight, nine, ten; ten." She was asked how many marbles were now in both cups. She immediately said "eighteen" and explained that "because there were eight and seven, and you know that eight and seven is fifteen, and then you put the other three, and there's fifteen; sixteen, seventeen, eighteen."

To count "fifteen; sixteen, seventeen, eighteen," Maya would have to establish "eight, nine, ten" first as *counted* items (three more than seven), and then reestablish them as *countable* items to be counted further "sixteen, seventeen, eighteen." This dual function of counted and countable items indicates that she could use her number sequence as its own material for creating countable items. Taking the results of counting as a situation of counting was manifest as a coordination between two number sequences—"fifteen; sixteen, seventeen, eighteen" and "seven; eight, nine, ten." It is this coordination (double counting) that is essential in children's construction of units-coordinating schemes.

Two-for-One Units-Coordinating Scheme

Born (1968) once commented that "the essential feature of mathematics is not numbers but the idea of coordination" (p. 176). Although Born's claim could be open to debate, it does fit our observations of what children do to establish "situations" that might be called "multiplicative." (The quotation marks indicate that it is the child's assimilated situation that the child construes as multiplicative.) For example, in the first teaching episode of the teaching experiment, a piece of red construction paper was placed in front of Maya together with several blue pieces cut so that six of them fit exactly on the red piece. Maya then was asked to find out how many blue pieces would fit on the red piece. After she said that six of them would fit, the blue pieces were removed and two orange triangles were placed on one blue piece (they fit exactly). Maya then was asked to find out how many orange pieces would go on the red piece without actually making the placements—"figure it out using the blue ones."

After looking straight ahead, she subvocally uttered number words and said "twelve." In explanation, she tapped the table twice with each of six fingers synchronous with uttering number words "one, two; three, four; five, six; seven, eight; nine, ten; eleven, twelve."

Tapping the table twice with each of six fingers indicates that Maya interpreted the request to "figure it out using the blue ones" numerically. But the counting activity was not the coordination that produced the "six twos." Rather, the coordination occurred prior to her counting activity. Because Maya could use her number sequence as its own material, she could create six units, each containing two units. The coordination involved inserting a unit of two into each of six units of one. That is, she "ran through" the six units symbolized by "six," inserting a unit of two into each unit of one.

This explanation is plausible because Maya had just counted the six blue pieces and "six" referred to the counted items. Being able to take counted items as "input" for further operating using her number sequence—as countable items—Maya could use her number sequence "one, two, three, four, five, six," established as a result of counting, as material on which to operate. She also used her established number sequence for "six" to guide and control the insertion of units of two into the six units of one. We believe that this second function of Maya's number sequence was not available to her consciousness. However, it is essential in our explanation because we have to account for Maya's forward distribution of a unit of two across six units of one. In this account, we believe that Maya would need to use her number sequence as an operator as well as an operand. This double use of her number sequence was not the novelty. Rather, inserting a unit of two into each element of her number sequence for "six" during the coordinating activity was the novelty.

The coordination that Maya made prior to counting constitutes what we call her assimilating operation. Through using her assimilating operation, she created her own "situation"—six twos—by introducing the novelty of distributing a unit of two across the elements of a unit of six. This multiplicative "situation," together with her goal of finding out how many orange pieces would fit on a red piece of paper, activated the activity of counting (which in this case was the activity of her multiplying scheme). The result of counting, when taken in conjunction with the situation and the activity, is what we call a "two-for-one units-coordinating scheme" (see von Glasersfeld, 1980, for a more complete analysis of schemes).

Being able to explain Maya's modification of her number sequence in order to construct her first concept of multiplication constituted a modification in our concepts of multiplication. We came to see multiplication as much different from a problem of changing a system of units, or a problem of repeated addition. Neither of these views of multiplication was rich enough to include Maya's understanding of multiplication. As happened with Confrey's (1990) concepts of exponential functions, our concepts of multiplication were transformed, elucidated, and enriched and now include this initial understanding of Maya's multiplying scheme.

Failure to Make a Units-Coordination in Re-Presentation

Maya's two-for-one units-coordination produced six twos—an experientially bounded plurality of units of two of numerosity six. The production of six twos was *a result* of the coordination. In the production of this structure, there is no reason to believe that Maya intentionally used her concept of two in imaginary experience, producing a bounded plurality of units of two. That is, there is no reason to believe that Maya used her concept of two to imagine a first unit of two, some intermediate units of two, and then a last unit of two that could be coordinated with a number sequence. At the very minimum, this is what Davydov's (1991) concept of multiplication involves—producing a units-coordination in re-presentation *prior to action.*

In order to investigate whether Maya could use her concept of two intentionally to create a bounded plurality of twos, the investigator asked Maya to make four rows of blocks with three blocks in each row. The rows were covered and Maya was asked how many blocks were in the rows. As Maya often did when she solved problems, she looked straight ahead in deep concentration and then said "thirteen." The investigator replied that she was off by one, so Maya repeated her solution process and said "twelve."

Feeling encouraged, the experimenter-teacher asked Maya to make two more rows like those she already had made and to place them under the cover with the original four rows. Maya then found how many rows were covered and then said that there would be 18 blocks because "you know you have twelve blocks and you add two more (rows), and you know two more is six, and you add six to go with the twelve!" Clearly, Maya could imagine two rows of three and operate with these rows, making and maintaining a distinction between the blocks in the rows and the rows. The question then arose in the teaching experiment whether this was sufficient for Maya to imagine a bounded plurality of rows of three of unknown numerosity. So the experimenter-teacher asked Maya how many rows of three she would need to add to the six rows to make 10 rows. After sitting silently for about 15 seconds, Maya said that she did not know and the experimenter-teacher reposed the question. Maya replied, "You mean to add three blocks to them under there?" The experimenter-teacher reaffirmed that *rows* of three were to be added, and again asked the question. Maya said, "One row and one more" and explained that "four more rows—you have six, and you count up to six and then take another three [blocks] and that's nine, and then one more and that's ten!"

Had the units been ones rather than rows of three, Maya could have used her number sequence in re-presentation (a regeneration of a past experience), creating a bounded plurality of number words starting with "seven" and ending with "ten" of unknown numerosity. She then could count, "seven is one; eight is two; nine is three; ten is four; four," to make the unknown numerosity known. This number sequence is what she used when trying to find how many rows were missing, and we see a conflation of units of three and units of one. She could not use the results of her units-coordinating scheme as its own input as she could use her number sequence.

When working with a known numerosity of rows, there was no necessity for Maya to use her concept of three in re-presenting a bounded plurality of rows of unknown number. In the case of four rows, she could use her concept of four in two ways when making four units of three— as supplying the units into which units of three were to be inserted and as a mechanism she could use to keep track of inserting the units of three. In the missing-row case, she could not make such a coordination because the number of times that she was going to use her concept of three to imagine a row was not specified. Her units-coordinations were restricted to known numbers of composite units and any notion of "times as many" that she had was restricted in the same way. It was even more limited, as we show in the next section.

Using a Composite Unit of Three in Iterating

Our understanding of the mathematics of students includes modifications in the schemes that we are able to bring forth in teaching episodes. Maya's units-coordinating scheme emerged as she found out how many orange triangles would go on the red piece of paper, a situation posed by the experimenter-teacher. We don't know whether Maya already had constructed the scheme prior to solving the situation. At the time of the actual teaching episode, this was not a meaningful question for the experimenter-researcher because he had no concept of Maya's units-coordinating scheme in the way we have explained. This explanation was formulated in a retrospective analysis. The situation was posed based on the experimenter-teacher's concept of multiplication, and he had no idea

that Maya would establish the "situation" that she did. In a very fundamental sense, the experimenter-teacher had set out to learn all that he
could about Maya's ways and means of operating in the context of interactive mathematical communication.

The experimenter-teacher learned that Maya could use her concept
of three in re-presentation and create a visualized image of three items.
This was indicated by how she used her concept of three when asked to
solve the missing-row situation above and by how she combined two
rows of three with four rows of three. In fact, Maya could imagine a pair
of threes. Because she conflated units of three and units of one when
trying to find how many more rows of three had to be added to six rows
to make ten rows, the experimenter-teacher hypothesized that Maya
could not use her concept of three in re-presentation to create a bounded
plurality of rows of three in visualized imagination—a first row of three,
some intermediate rows, and then a last row of three. Maya could imagine
two rows of three, but that is not the same thing as imagining a bounded
plurality of rows of three because visual patterns can be used to imagine
two rows of three. To make progress, Maya would have to modify her use
of her concept of three. She had to shift from using the three just once or
twice to using it an unknown number of times.

In a situation in which Maya made seven rows of three, the experimenter-teacher hid three more rows under a cover, told her that there
were now 10 rows in all, and asked her how many rows of three were
covered. After looking at the visible rows for about 45 seconds, Maya said
"three rows" and explained, "because you have seven, and another row
of three equals eight, and another row of three equals nine, and another
row of three equals 10!" To check her operating, the experimenter-teacher
asked Maya how many blocks were hidden. Maya again looked straight
ahead before saying "nine."

Maya definitely made a modification in her units-coordinating
scheme in that, for the first time, she could coordinate sequentially producing her concept of three with producing the elements of her number
sequence from seven up to and including 10. Having the seven rows visible seemed to be crucial for her modification. She was able to imagine
this series of rows continuing beneath the cover without deliberately using her concept of three in re-presentation more than once. Projecting the
seven visible rows as continuing on under the cover provided a situation
in which she could re-present sequentially rows of three. As she was sequentially imagining the rows, she coordinated this productive activity
with a production of her number sequence: "because you have seven, and
another row of three equals eight, and another row of three equals nine,
and another row of three equals ten!" We view the coordination of these

two productive activities as a process of externalizing the units-coordination. She now could use her concept of three in iterating and could combine the result of each iteration with those of the preceding. Correspondingly, we call her unit of three an iterating unit.

Maya's units-coordinating scheme still was restricted to a specific number of composite units of equal and known numerosity. But she had made progress in that she could produce an enactively and sequentially unknown number of composite units and combine this productive activity with a production of her number sequence. However, we still had no indication that Maya could produce a bounded plurality of composite units *in re-presentation* prior to producing them enactively and sequentially. It seemed that the seven visible rows of three were necessary for her to engage in iterating her unit of three.

There was a fundamental distinction between the situations that Maya could establish as multiplicative and those she couldn't. Whenever she knew the number of composite units involved (say, four rows) as well as the numerosity of each composite unit (say, three blocks per row) she could make a units-coordination and find the numerosity of the elements of the composite units (the number of blocks). However, when the numerosity of the elements of a composite unit was given along with the numerosity of the elements of another composite unit (How many rows of three can you make using nine blocks?), a units-coordination was made only enactively if she was asked to find how many of the first she could make using the elements of the second. That is, she could *segment* enactively the second composite unit using the first, but she could not produce a re-presentation of the results of segmenting prior to action.

DISCUSSION

By engaging in the teaching experiment, we learned that Maya had not constructed multiplication as defined by Davydov (1991). To produce the *insight* that a box of coins can be weighed to find out how many coins are in the box, one has to imagine a composite unit of coins of unknown numerosity weighing, for example, one kilogram. One then has to imagine a bounded plurality of such composite units—a first such composite unit, some intermediate composite units, and then a last composite unit—that can be made from the coins in the box, also of unknown numerosity. We could see no indication in the teaching experiment that Maya used her concept of three in re-presentation *prior to acting* to establish a bounded plurality of threes in visualized imagination. In this, there is a necessity for these composite units to be *countable*. Otherwise, there would be no possibility for finding how many composite units of three could be made using nine blocks.

The imagined plurality of composite units together with their countability creates a multiplicative structure that is imputed to the situation by the cognizing individual. Creating such a multiplicative structure was far beyond Maya's more primitive multiplying scheme. For this reason, we feel fully justified in differentiating Maya's schemes and Davydov's (1991) concept of multiplication. In fact, multiplication cannot be "taught" if that term refers to transporting our adult concepts to children. Rather, children contribute the assimilating operations of coordinating, and on that basis produce their own ways and means of operating that adults might call "multiplication."

Maya's units-coordinating scheme could be said to be implicit in her number sequence because it was produced as a modification of double counting. For this reason, we refer to her complete units-coordinating scheme as an *implicit concept of multiplication*. This choice of terminology is justified because she had only begun the process of externalizing the units-coordination when establishing her unit of three as an iterating unit in attempting to solve a missing-row situation.

It seemed to be natural for Maya to combine four rows of three and two rows of three to find how many blocks were in the six rows by adding the blocks in the two rows to those in the four rows. But this should not be interpreted as repeated addition in the sense of taking three as six addends. A units-coordination scheme, as we have explained it, is a root structure that can yield what one might want to call "repeated addition," but it should not be confused with repeated addition in contemporary school mathematics. Maya's units-coordinating scheme taught us how to begin to make a distinction between children's meanings of addition and multiplication.

CONCLUSION

Although the teaching experiment described here served primarily as a research tool to gain understanding of children's mathematics and to help us to reconceptualize our own understanding of mathematics, we believe that the findings have significant implications for teaching. An interpretation of mathematics through the students' voices is critical to the successful teaching of the discipline. Through the voice of Maya, we have learned that the multiplying scheme she used differed greatly from the ones that we might have expected her to use. Based on the information learned from this experience, coupled with what has been learned in other research experiences, we are convinced that there is a need to rethink the approach used to introduce multiplication in schools. The evidence suggests that a much stronger link to children's numerical schemes would be a more effective approach than what currently is used in our elementary curricula.

Ultimately the mathematics curriculum is shaped by the teacher's knowledge of students' mathematics. As constructivists, we believe that teachers' knowledge of students' mathematics must be built up as the teachers engage in interactions with students because they do not have the luxury of retrospective analysis of videotaped teaching episodes. The nature of these interactions often determines the degree of insight that a teacher has about the students' understanding. Our own experience in conducting teaching experiments has revealed teaching experiments as a powerful tool to access students' learning in action. Although it is unreasonable to use teaching experiments with each individual student on a daily basis in a regular classroom, there are several elements of the teaching experiment that warrant consideration by all teachers.

The most striking element is the insight that teachers can gain by interacting with students in ways that allow the teachers to explore the students' ways of thinking and operating. Teachers who take the time to listen carefully to the students will realize that the mathematics of students is the most powerful input that teachers can use to shape their teaching practice. However, the mathematics of students must be brought forth, sustained, and modified by teachers. It is our intention to help teachers to understand the mathematics of their students by providing explanations of students' schemes of action and operation that we have been able to co-create with students.

ACKNOWLEDGMENTS

We would like to thank the editors of this volume and Dr. Heide Wiegel for their comments on an earlier version of this paper.

REFERENCES

Born, M. (1968). *My life and my views*. New York: Charles Scribner's Sons.

Cobb, P., & Steffe, L. P. (1983). The constructivist researcher as teacher and model builder. *Journal for Research in Mathematics Education, 14*(2), 83–94.

Commission on Standards for School Mathematics. (1989). *Curriculum and evaluation standards for school mathematics*. Reston, VA: National Council of Teachers of Mathematics.

Confrey, J. (1990). The concept of exponential functions. In L. P. Steffe (Ed.), *Epistemological foundations of mathematical experience* (pp. 124–159). New York: Springer-Verlag.

Davydov, V. V. (1991). A psychological analysis of the operation of multiplication. In L. P. Steffe (Vol. Ed., English-language ed.), *Psychological abilities of primary school children in learning mathematics* (pp. 9–85). Soviet Studies in Mathematics Education, Vol. 6. Reston, VA: National Council of Teachers of Mathematics.

Hunting, R. (1983). Emerging methodologies for understanding internal processes governing children's mathematical behaviour. *The Australian Journal of Education, 27*, 45–61.

Sinclair, H. (1988). Foreward. In L. P. Steffe & P. Cobb (Eds.), *Construction of arithmetical meanings and strategies* (pp. v–vi). New York: Springer-Verlag.

Steffe, L. P. (1983). The teaching experiment methodology in a constructivist research program. In M. Zweng et al. (Eds.), *Proceedings of the Fourth International Congress on Mathematical Education* (pp. 469–471). Boston: Birkhauser.

Steffe, L. P. (1992). Schemes of action and operation involving composite units. *Learning and Individual Differences, 4*, 259–309.

Steier, F. (1995). From universing to conversing: An ecological constructionist approach to learning and multiple description. In L. P. Steffe & J. Gale (Eds.), *Constructivism in education* (pp. 67–84). Hillsdale, NJ: Erlbaum.

von Glasersfeld, E. (1980). The concept of knowledge in a constructivist theory of knowledge. In F. Benseler, P. M. Hejl, & W. K. Kock (Eds.), *Autopoiesis, communication, and society: A theory of autopoietic systems in the social sciences* (pp. 75–85). New York: Campus-Verlag.

von Glasersfeld, E. (1989). Constructivism in education. In T. Husén & N. Postlethwaite (Eds.), *International encyclopedia of education* (Supplementary Volume 1, pp. 162–163). Oxford, England: Pergamon.

PART II

Improving Curriculum and Teaching in Science and Mathematics

CHAPTER 7

Reorganizing the Curriculum and Teaching to Improve Learning in Science and Mathematics

Reinders Duit
University of Kiel, Germany

Jere Confrey
Cornell University, Ithaca, New York, United States

The constructivist view has important consequences for the development of new teaching and learning approaches that focus on students' understanding of science and mathematics rather than recall of facts and formulas. The scope of the constructivist approaches presented in the literature is very broad. On the one hand, there are studies that involve new teaching and learning conditions only on a "micro-level" (e.g., the impact of a particular teaching method or teaching aid such as a new computer program). On the other hand, there are long-term projects to develop new approaches that include a wide range of topics and methods. Serious constructivist approaches usually set out to reorganize traditional teaching by including changes of aims, setup of content structures, media, and teaching/learning strategies.

In science, Driver and Scott (see Chapter 8, this volume) introduce constructivist ideas by changing traditional teaching and learning in many respects. In mathematics, Maher and Alston (1990) proposed a long-term project that involves teachers in constructivist reform efforts. In describing the implications for classroom teaching, they focus on three issues that arise repeatedly: how to learn to listen to students' thinking; how to organize classroom activities to support "listening and questioning"; and how to implement forms of assessment that document children's questions. They focus on the importance of long-term learning. Both projects are paradigmatic for constructivist attempts in that the roles of teachers and researchers are fundamentally different from those in traditional curriculum designs. Constructivist curriculum development rejects the traditional curriculum model and replaces the strong dominance of the researchers by a model in which teachers and researchers are equal partners in the curriculum process (Driver & Oldham, 1986).

With these ideas in mind, in this chapter we discuss pertinent aspects of conceptual growth and conceptual change, describe five assumptions for reorganizing curricula and teaching from a constructivist perspective, and provide examples of the development of new media and alternative teaching strategies for reorganizing the curricula and teaching to improve learning in science and mathematics.

CONCEPTUAL GROWTH AND CONCEPTUAL CHANGE

Two Complementary Kinds of Learning

Distinctions have been made in science and mathematics education between conceptual growth and conceptual change. "Conceptual growth" refers to enlargement of the conceptual network in such a way that one's previous knowledge and its connections, for the most part, remain intact. In science, this can refer to the addition of new terminology, definitions, and isolated facts that do not disturb one's previous beliefs, but only add to the existing knowledge base. In mathematics, conceptual growth would include the development of skills, techniques, and examples within a conceptual framework. For instance, learning the algorithm for multiplication when one has learned to multiply with Dienes blocks might be an example of conceptual growth for most students. (Of course, it could involve conceptual change if it produced a reconsideration of place value issues in the two numbers.) Conceptual growth is akin to the evolutionary growth of science through the conduct of "normal" science (Kuhn, 1970). In Piaget's framework, conceptual growth is like the idea of assimilation.

Accommodation, in Piaget's terminology, is comparable to conceptual change (Dykstra, 1992; Petrie, 1979). The term "conceptual change" evolved from Toulmin's (1972) claim in the philosophy of science that understanding the path of science is undertaken best by examining the periods of conceptual change. For Kuhn (1970), a revolutionary change in the conduct of science such as a paradigm shift marks a discontinuity in the conduct of the discipline. Lakatos (1976) argued for a slightly different approach to the evolution of mathematical thought by suggesting that mathematics proceeds as a dialectic of proofs and refutations, so that conceptual change occurs in fits and starts as conjectures are proposed, explored, and refuted or revised.

In education, "conceptual change" refers to cognitive restructuring different from what is evidenced in conceptual growth. Some researchers note that conceptual change also involves an underlying continuum that can vary from weak to strong restructuring. Whether learning something requires conceptual change or weak restructuring is largely a function of the preconceptions of the learner. However, researchers have documented convincingly that certain topics are more likely than others to result in one form of learning or the other.

Conditions That Support Conceptual Change

The term "conceptual change" has been used in manifold meanings in science and mathematics education research on learning and instruction. But there appears to be a common concern that this term stands for a kind of learning where a major restructuring of the already existing conceptual structure is necessary as outlined above. The initial theory of conceptual change (Posner, Strike, Hewson, & Gertzog, 1982) that may be indicated by the four conditions of dissatisfaction-intelligible-plausible-fruitful (see Hewson, Chapter 11, this volume; also remarks on the Conceptual Change Model below) has been discussed critically (Pintrich, Marx, & Boyle, 1993; Strike & Posner, 1992; contributions in Vosniadou, 1994). Very briefly summarized, it is claimed that considerations concerning pathways from students' preinstructional conceptions toward the science or mathematics conceptions should not put emphasis mainly on the logical plane based on the structure of the referring science and mathematics content. Conceptual change, in other words, may not be primarily viewed as a purely rational process but motivational beliefs and classroom contextual factors are of key significance in facilitating conceptual change (Pintrich et al., 1993). Conceptual change happens only if it is embedded in an appropriate set of what may be called conceptual change supporting conditions.

Change or Extinction of Students' Preconceptions?

Hewson (Chapter 11, this volume) explains that the word "change" can be used in different ways—it can mean the replacement of one idea by another, an increase or decrease in the amount of something, or something's gaining status while something else loses status. Hewson argues that the last explanation for change is best suited for describing learning science because it is neither possible nor wise to extinguish students' everyday conceptions, which have proven to be sustainable and viable in explaining everyday phenomena. A similar view is expressed by Grandy (1990), who argues that "extinguishing" preconceptions is a misnomer.

In mathematics education, the issue has developed in a related but distinct manner. Researchers have documented the hidden competencies of vendors, tailors, architects, carpenters, and so forth, outside the classroom setting, and contrasted these with weak performances by the same individuals on standardized measures (Lave & Wenger, 1991; Saxe, 1995). In doing so, they have demonstrated the role of context in promoting arithmetical and geometrical reasoning. For the most part, these informants describe themselves as inept in mathematics, and identify it with school algorithms. A striking discontinuity is apparent between the informal mathematics of the consumer and user of mathematics and the formal treatment of the same topics in school. Proponents of reform argue for more frequent inclusion of situational variables in classroom mathematics instruction, rather than the extinction of the existing competencies. Thus, there is a growing recognition that the preconceptions of students must become a part of the classroom discourse and that the content to be "taught" should undergo reform to incorporate these perspectives. Premature introduction of formal algorithms can cause students to undermine their intuitive base and to learn the algorithms in a rote or mechanical fashion.

A second debate in mathematics is related to conceptual change. Often in mathematics, an informally held belief ("primitive model" in the terms of Fischbein, Deri, Nello, & Marino, 1985) functions effectively within a limited domain (e.g., interpreting multiplication as repeated addition, and therefore assuming that multiplication makes numbers larger). However, when one extends the domains (Hawkins, 1974) to include multiplication by decimals less than one, this intuitive view fails and the answers become counterintuitive for the students. In mathematics, as in science, some argue that such an extension must be made on formal grounds and that the denial of the intuitive is part of developing a mature disciplinary perspective. The conceptual change literature suggests that the transition to a more formal conception will be successful only if recognition of the viability of the initial conceptions is made, and the new approaches provide new possibilities for problem-solving progress. Furthermore, there is an epistemological challenge in deciding what is intended by a formal approach and how this approach can be changed to integrate more closely with student conceptions.

Continuous or Discontinuous Passage from Students' Preconceptions to Science and Mathematics Conceptions?

Research has shown that learning science and mathematics frequently requires the process of conceptual change, more than the process of conceptual growth, or that conceptual growth occurs only during the period following conceptual change when new connections can be made. As a result, many of the constructivist teaching and learning approaches focus on this kind of learning. Some approaches (such as those in Driver and Scott, Chapter 8, this volume) start by eliciting students' alternative conceptions, and then involve some sort of cognitive conflict to draw attention to the differences between students' conceptions and those of science. Such approaches involve viewing conceptual change mainly as a discontinuous passage. Other approaches try to avoid conceptual change of this kind by starting with aspects of students' intuitive conceptions that are mainly in accordance with the official scientific views, and then work toward enlargement of these ideas through "partial restructuring" (Brown & Clement, 1989; Grayson, Chapter 13, this volume; Stavy, 1991). Such strategies are undoubtedly valuable provided that it is possible to find a fruitful starting point that allows a continuous passage. In science, this often appears to be impossible to achieve and, hence, approaches involving a discontinuous passage are necessary.

In mathematics, it is important not to confuse conceptual growth with the accumulation of skills in the absence of a conceptual framework. The difference between procedural and conceptual understanding is often used to make this distinction. For example, Hiebert and Lefevre (1986) describe procedural knowledge as symbolic representational systems, consisting of syntactic conventions, algorithms, procedures, and problem-solving strategies and operations, whereas conceptual knowledge involves interrelationships among concepts. This is reminiscent of the distinction made by Skemp (1978) between instrumental and relational knowledge. Both of these researchers emphasize that procedural knowledge becomes an effective mathematical tool only within the framework of conceptual understanding.

Cognitive Conflict

In approaches that focus on conceptual change, cognitive conflict strategies play a key role (Scott, Asoko, & Driver, 1992). Cognitive conflict can be created by asking for students' predictions and then contrasting these with the experimental results, by contrasting the ideas of the students and those of the teacher, and by contrasting the beliefs among the students.

Cognitive conflict seems to play an important role in constructivist instruction. Theoretically, it underlies Piaget's genetic epistemology in which disequilibration demands an interplay between assimilation and accommodation until equilibration is restored (Dykstra, 1992; Rowell & Dawson, 1985) and frequent reference is made to Festinger's theory of cognitive dissonance (Driver & Erickson, 1983). The crucial point in instruction, however, is whether students really "see" the conflict. Sometimes what is clearly discrepant or contradictory to the teacher (or expert) either is not discrepant at all or is only marginally different from the student's point of view. For example, Balacheff (1991) documents the complexity in having students recognize a counterexample in the context of conjecturing about the number of diagonals in a polygon. Studies such as this do not suggest the necessity of discarding theories of cognitive conflict, but they do imply that arranging a potential conflict can be difficult and that the role of the teacher in assisting in the process is essential (Bauersfeld, 1991; Cobb, Wood, & Yackel, 1991). Conceptual-change ideas can be contrasted with much of the problem-solving literature in terms of whose perspective is used in defining a cognitive conflict. Whereas the problem-solving literature tends to locate a problem within the ontology of mathematical or science objects, a constructivist thinks of a problem in relation to the student who is working on it (Confrey, 1991, p. 117).

RETHINKING THE AIMS OF SCIENCE AND MATHEMATICS INSTRUCTION

There appear to be five assumptions shared by mathematics and science educators for reorganizing the curriculum and teaching to improve learning in school science and mathematics from a constructivist perspective.

First, constructivist approaches usually give more emphasis to the applicability of science and mathematics knowledge in situations in which students are interested than do more traditional approaches. Constructivist approaches aim at science and mathematics knowledge not as stored items in memory or magical formulas but as "knowledge in action" (Driver & Erickson, 1983), "tools for modelling" (Nemirovsky & Rubin, 1991), "situated learning" (Lave & Wenger, 1991), or "authentic learning situations" (Roth, 1995). The mathematics and science of daily life become appropriate content for building closer connections between prior knowledge and scientific or mathematical content.

Second, constructivist approaches demand the introduction into the curriculum of issues of *meta-knowledge* about science and mathematics (Ball, 1990; Niedderer, 1987; Thompson, 1994). Students have to learn not only mathematics and science content, but also basic ideas about the nature and range of mathematics and science principles and concepts. For example, the relational view of knowledge that is characteristic of constructivism has to be given emphasis. Students should learn that mathematics and science knowledge is a tentative human construction and not an eternal truth.

Third, the implementation of constructivist approaches in science and mathematics has demonstrated the impossibility and inadvisability of extinguishing students' everyday conceptions while replacing them with "scientific" or "mathematical" knowledge as outlined above. This "replacement" view must be revised to allow for the coexistence and articulation of a relationship between informal and formal conceptions. More elaborate and complex mathematical and scientific ideas must be demonstrated to allow deeper or broader explanations if they are to become viable to the students. In mathematics especially there is an increasing emphasis of the construction, interpretation, and coordination of multiple representations of ideas. This has led to a focus on how students form connections among ideas and recognize patterns of variation and invariance (Janvier, 1983; Rubin, 1990).

Fourth, constructivist approaches are student-centered. Of course, most other approaches make this claim also. However, constructivist approaches are student-centered in a specific way—they use subject matter as a vehicle for interactive engagement with students. Ideas are embedded in student-oriented challenges and the classroom climate supports and encourages active exchange, debate, and negotiation of ideas. These challenges seek to alter not only students' narrow views of the scientific and mathematical enterprise, but their assumptions about expertise, learning, and teaching. They do not attempt to make learning simple or entertaining, but place a high demand on students (and teachers) to engage with the ideas in a deep and sustained manner. There is considerable evidence that, at first, students will be perplexed and will even resist such instruction, because they have become relatively complacent, disengaged, and "pleased" with methods that allow them to learn pieces of knowledge by heart (White, 1986). To achieve a change in students' explanations of classroom participation, researchers have found it essential to promote the development of students' sense of autonomy (Confrey, 1991). Students must be weaned from their reliance on teachers' assessment of progress and success, and this requires students to become more keenly aware of their own thinking processes. To this end, methods to enhance reflection on student methods and forms of argument become commonplace in the constructivist classroom.

Fifth, constructivist researchers recently have recognized that the norms, routines, and patterns of classroom interaction form a fundamental influence on the effectiveness of reform efforts (Bauersfeld, 1991; Cobb et al., 1991; Voigt, 1985). Students' alternative conceptions of mathematics and science are stabilized when conventions are taken as logical necessity, and the role of the community in establishing those conventions is minimized. Successful constructivist approaches are based on revising classroom norms to value alternative perspectives, thus reaching negotiated consensus about how classes are to be conducted and the possible meanings of and agreements about the content being taught.

IMPLEMENTING THE NEW AIMS OF SCIENCE AND MATHEMATICS INSTRUCTION

The five assumptions described above for reorganizing the curriculum and teaching to improve science and mathematics from a constructivist perspective can be implemented in a variety of ways. In this section, we describe this implementation process in terms of developing new media, revising traditional content structures, and using a range of constructivist teaching strategies.

Development of New Media

The development of new teaching aids such as experiments, textbooks, and other media has always been a major part of science and mathematics education research and evaluation. There also have been several attempts to develop new media from a constructivist perspective. What appears to be different in these attempts compared with more traditional ones is that they do not try to improve learning just by providing one new teaching aid; instead, each new item usually is integrated into a comprehensive constructivist approach.

The computer, in particular, is seen as a promising tool. The implications of the introduction of computer technology into the curriculum are far-reaching. Not only can some computer environments provide students with relief from some of the drudgery of repetitive computation, but they can open up possibilities that previously could be made accessible only through the imagination (Confrey & Smith, 1991; Kaput, 1987). A very promising use of the computer within the constructivist perspective seems to be as a tool for exploring problem solutions. The computer can provide opportunities for dynamic displays and visualizations, simulations, and model-building (Goldberg & Bendall, Chapter 5, this volume). These capacities can have a profound impact on the content to be learned and understood. Not only do students' preconceptions become of great importance, but so too do their forms of inquiry because students must learn how to move about in a different intellectual space. In mathematics, Goldenberg and colleagues (1988) have demonstrated that the rich resources offered by the computer bring their own set of issues concerning what students perceive and their attributional patterns.

The textbook continues to be a major source of information and is a major starting point for learning science and mathematics (Yore, 1991). It is somewhat surprising, therefore, that so little attention has been devoted to this medium by constructivist researchers in science and mathematics. Textbooks appear to be viewed as unsuitable for constructivist teaching and learning approaches. Recent critiques of science textbooks have revealed that the books used today are based mainly on traditional nonconstructivist ideas in science. Usually, the learner's perspective is not taken into consideration, and issues of the psychology of learning are neglected (Stinner, 1992). Furthermore, textbooks usually are written in a limited authorial style (Strube, 1989) and they provide limited empiricist views of the nature of science (Sutton, 1989). There is no doubt that the latter two critiques could be addressed in textbooks. Where learning issues are concerned, there also appear to be promising ways of improving textbooks from the constructivist perspective (Duit, Haeussler, Lauterbach, Mikelskis, & Westphal, 1992). Guzetti, Snyder, Glass, and Gamas's (1993) meta-analysis of studies suggested that conceptual change can be promoted by the use of science textbooks provided that students' preconceptions are challenged either by text that is refutational or text used in combination with other strategies that cause cognitive conflict.

Revising Traditional Content Structures

Another active field of science and mathematics education research involves the development of alternatives to the content structures previously used in schools (Fensham, Gunstone, & White, 1994). It is sometimes really surprising to find that a specific topic (for example, energy) can be introduced and developed in many different ways and that very different content structures appear to be valuable. There are several approaches in the field of constructivist science and mathematics teaching that now involve the design of new content structures in order to avoid traditional misunderstandings. Brown and Clement's (1992) constructivist approach to Newtonian mechanics was successful only when the content structure of the first approach also was revised significantly in such a way that students' learning difficulties could be addressed more appropriately.

In particular, approaches that involve a continuous passage from students' conceptions to science and mathematics conceptions provide alternative content structures. Duit and Haeussler (1994) argue in favor of an approach to the concept of energy in science that starts from students' everyday ideas of energy. Those everyday ideas that are not in accordance with the science view are reinterpreted in such a way that, on the one hand, they are still valid in everyday use but, on the other, they achieve a sound science underpinning. The traditional path to the energy concept via the "ability to do work" has no place in such an approach.

In mathematics, constructivist investigations have led to significant reorganization of the teaching of such topics as counting (Steffe, von Glasersfeld, Richards, & Cobb, 1983), multiplication (Steffe & D'Ambrosio, Chapter 6, this volume), and the construction of units (Steffe, 1994), rational number (Kieren, 1976), functions (Confrey & Doerr, Chapter 14, this volume; Dreyfus & Eisenburg, 1984), and calculus concepts (Nemirovsky, Mokros, & Barclay, 1990).

Constructivist Teaching Strategies

Constructivist teaching strategies comprise the area that is given most attention in research. As is outlined above, the constructivist view demands very specific ways of teaching and learning. A comprehensive review of different strategies is given by Scott et al. (1992). Below, key approaches on a macro level are reviewed. Single-intervention methods, such as the use of analogies and metaphors (see the review by Duit, 1991) or Socratic dialogue, are considered only if they are part of a broader approach.

Driver's *constructivist teaching sequence* (see Driver & Scott, Chapter 8, this volume) is paradigmatic for many other attempts to take students' preinstructional conceptions into consideration in a serious way. In such approaches, there is always a phase in which students' ideas are elicited and discussed in class before the science or mathematics view is presented. A key feature in the phase of contrasting students' ideas and the science and mathematics conceptions is negotiation in which the teacher is a facilitator rather than a transmitter of knowledge. However, students are often unable to see the differences between their views and the science and mathematics view introduced by the teacher. Further, younger students do not like to play around with ideas—they want to know the "right" ("true") answer (Baird & Mitchell, 1986).

The *learning cycle approach,* which originates from a Piagetian perspective, recently has been proposed explicitly as a means to enhance conceptual change (Stepans, Dyche, & Beiswenger, 1988). There are three phases in this strategy, namely, *exploration, term introduction,* and *concept application.* The exploration phase can be compared to the orientation and eliciting phase in Driver and Scott's (Chapter 8, this volume) model, but the emphasis here is slightly more on students' having experiences with new phenomena than on students' exchange of ideas about the phenomena. Cognitive conflict plays a major role in both teaching strategies.

At the center of the *conceptual change model* (see Hewson, Chapter 11, this volume) are the four conditions of conceptual change. There must be *dissatisfaction* with existing ideas, and the new conception must be *intelligible, initially plausible,* and *fruitful.* The first and the last conditions have proven to be the most difficult for teachers to arrange. Students often are pleased with their everyday conceptions and experience no dissatisfaction. Further, it is difficult to persuade students that the new conception is more fruitful than their everyday conceptions. The conditions of being intelligible, plausible, and fruitful are indicators of what Hewson calls the "status" of a conception. Hewson argues for a change of status in students' conceptions and not for extinction of the "old" everyday preconceptions. Students have to be convinced that, in certain situations, science and mathematics conceptions are more intelligible, plausible, and fruitful than their everyday conceptions.

Brown and Clement (1989) have developed a *bridging analogies approach* that tries to find a largely continuous passage toward the science view from facets of students' conceptions that are mainly in accordance with the science view already. The key idea is to facilitate this passage by a series of "stepping-stones" that are designed as bridging analogies. A paradigmatic example of bridging analogies is explaining the situation in which "a book is lying on a table." Students have difficulty not only with the idea that the book exerts a force on the table, but also with the idea that the table "pushes back" (i.e., the table exerts a force on the book). As the table is not active, students do not consider that there is a force on the book caused by the table. To guide students toward an understanding of this situation, bridging situations are designed. The starting point is the situation in which a spring is compressed by a finger. In this case, Brown and Clement (1989) think that the correct intuition is triggered by the idea that the spring "pushes back."

CONCLUSION

Providing evidence of success of constructivist approaches is somewhat difficult. It is not easy to condense promising results into measures that allow comparisons with traditional approaches. As constructivist approaches usually aim at a fundamental restructuring of traditional settings, there is the problem of what to base the measures on. Do categories have to come from the traditional approaches or from the new constructivist ones? There is also the issue that it is not only understanding of science and mathematics content that matters in constructivist approaches, but also issues of a satisfactory classroom climate. Nevertheless, research results that allow comparisons indicate that conceptual-change approaches are mostly superior to traditional approaches where understanding science and mathematics content is concerned.

Research involving student conceptions can offer a significant amount of guidance for the improvement of curriculum and teaching in science and mathematics by encouraging the articulation and development of student conceptions. The developmental process includes processes of assimilation and accommodation, and allows for closer connections between children's experiential world and disciplinary knowledge. Increasingly, the responsibility for such development lies with an autonomous and reflective student who learns to express her or his beliefs, investigate them in the company of other students and teachers, and learn to internalize more robust ideas about science and mathematics. The classrooms that will be most effective in reforming their practices will be ones in which students' constructive activities are supported by the interlocking system of patterns and routines that promote communication, reflection, and the effective use of resources.

REFERENCES

Baird, J. R., & Mitchell, I. J. (1986). *Improving the quality of teaching and learning—An Australian case study.* Melbourne, Australia: Monash University.

Balacheff, N. (1991). Treatment of refutations: Aspects of the complexity of a constructivist approach to mathematics learning. In E. von Glasersfeld (Ed.), *Radical constructivism in mathematics* (pp. 89–110). Dordrecht, The Netherlands: Kluwer.

Ball, D. (1990). *Halves, pieces and twoths: Constructing representation contexts in teaching fractions* (Craft Paper 90-2). Reston, VA: National Center for Teacher Education.

Bauersfeld, H. (1991). *Classroom cultures from a social constructivist's perspective* (Occasional Paper 131). Bielefeld, Germany: Institute for Mathematics Education, University of Bielefeld.

Brown, D. E., & Clement, J. (1989). Overcoming misconceptions via analogical reasoning: Abstract transfer versus explanatory model construction. *Instructional Science, 18,* 237–261.

Brown, D. E., & Clement, J. (1992). Classroom teaching experiment in mechanics. In R. Duit, F. Goldberg, & H. Niedderer (Eds.), *Research in physics learning: Theoretical issues and empirical studies* (pp. 380–397). Kiel, Germany: Institute for Science Education (IPN), University of Kiel.

Cobb, P., Wood, T., & Yackel, E. (1991). A constructivist approach to second grade mathematics. In E. von Glasersfeld (Ed.), *Radical constructivism in mathematics* (pp. 157–176). Dordrecht, The Netherlands: Kluwer.

Confrey, J. (1991). A student's understanding of the power of ten. In E. von Glasersfeld (Ed.), *Radical constructivism in mathematics* (pp. 111–138). Dordrecht, The Netherlands: Kluwer.

Confrey, J., & Smith, E. (1991). A framework for functions: Prototypes, multiple representations, and transformations. In R. Underhill & C. Brown (Eds.), *Proceedings of the 13th Annual Meeting of PME-NA* (pp. 57–63). Blacksburg, VA: Division of Curriculum and Instruction.

Dreyfus, T., & Eisenburg, T. (1984). Intuitions of functions. *Journal of Experimental Education, 52,* 77–85.

Driver, R., & Erickson, G. L. (1983). Theories-in-action: Some theoretical and empirical issues in the study of students' conceptual frameworks in science. *Studies in Science Education, 10,* 37–60.

Driver, R., & Oldham, V. (1986). A constructivist approach to curriculum development in science. *Studies in Science Education, 13,* 105–122.

Duit, R. (1991). On the role of analogies and metaphors in learning science. *Science Education, 75,* 649–672.

Duit, R., & Haeussler, P. (1994). Learning and teaching energy. In P. Fensham, R. Gunstone, & R. White (Eds.), *The content of science: A constructivist approach to its teaching and learning* (pp. 185–200). London, England: Falmer Press.

Duit, R., Haeussler, P., Lauterbach, R., Mikelskis, H., & Westphal, W. (1992). Combining issues of "girl-suited" science teaching, STS, and constructivism in a physics textbook. *Research in Science Education, 22,* 106–113.

Dykstra, D. J. (1992). Studying conceptual change: Constructing new understandings. In R. Duit, F. Goldberg, & H. Niedderer (Eds.), *Research in physics learning: Theoretical issues and empirical studies* (pp. 40–58). Kiel, Germany: Institute for Science Education (IPN), University of Kiel.

Fensham, P. J., Gunstone, R. F., & White, R. T. (1994). *The content of science: A constructivist approach to its teaching and learning.* London, England: Falmer Press.

Fischbein, E., Deri, M., Nello, M. S., & Marino, M. S. (1985). The role of implicit models in solving verbal problems in multiplication and division. *Journal for Research in Mathematics Education, 16,* 3–17.

Goldenberg, E. P., et al. (1988). *Mathematical, technical, and pedagogical challenges: In the graphical representation of functions* (Technical Report). Newton, MA: Center for Learning Technology, Education Development Center, Inc.

Grandy, R. E. (1990, April). *On the strategic use of history of science in education.* Paper presented at the meeting of the American Educational Research Association, Boston, MA.

Guzetti, B. J., Snyder, T. E., Glass, G. V., & Gamas, W. S. (1993). Promoting conceptual change in science: A comparative meta-analysis of instructional interventions from reading education and science education. *Reading Research Quarterly, 28,* 116–159.

Hawkins, D. (1974). *The informed vision.* New York: Agathon Press.

Hiebert, J., & Lefevre, P. (1986). Conceptual and procedural knowledge in mathematics: An introductory analysis. In J. Hiebert (Ed.), *Conceptual and procedural knowledge: The case of mathematics* (pp. 1–28). Hillsdale, NJ: Erlbaum.

Janvier, C. (1983). Representation and understanding: The notion of function as an example. In R. Hershkowitz (Ed.), *Proceedings of the Seventh International*

Conference for the Psychology of Mathematics Education (pp. 226–271). Rehovot, Israel: Weizmann Institute of Science.

Kaput, J. (1987). Representation systems and mathematics: Opening new representational windows. In C. Janvier (Ed.), *Problems of representation in the teaching and learning of mathematics* (pp. 19–26). Hillsdale, NJ: Erlbaum.

Kieren, T. (1976). On the mathematical, cognitive, and instructional foundations of rational numbers. In R. Lesh (Ed.), *Number and measurement* (pp. 101–144). Columbus: ERIC/SMEAC, Ohio State University.

Kuhn, T. S. (1970). *The structure of scientific revolutions.* Chicago: University of Chicago Press.

Lakatos, I. (1976). *Proofs and refutations.* Cambridge, England: Cambridge University Press.

Lave, J., & Wenger, E. (1991). *Situated learning: Legitimate peripheral participation.* Cambridge, England: Cambridge University Press.

Maher, C., & Alston, A. (1990). Teacher development in mathematics in a constructivist framework. In R. Davis, C. Maher, & N. Noddings (Eds.), *Constructivist views on the teaching and learning of mathematics* (pp. 147–166). Reston, VA: National Council of Teachers of Mathematics.

Nemirovsky, R., Mokros, F., & Barclay, W. (1990). *Measuring and modelling project: First year annual report* (Unpublished TERC report). Cambridge, MA: Technical Education Research Center.

Nemirovsky, R., & Rubin, A. (1991). "It makes sense if you think about how graphs work. But in reality . . . ." In F. Furinghetti (Ed.), *Proceedings of the Fifteenth PME Conference* (pp. 57–64). Assisi, Italy.

Niedderer, H. (1987). A teaching strategy based on students' alternative frameworks—Theoretical concept and examples. In J. Novak (Ed.), *Proceedings of the Second International Seminar on Misconceptions and Educational Strategies in Science and Mathematics* (Vol. II, pp. 360–367). Ithaca, NY: Cornell University.

Petrie, H. G. (1979). Metaphor and learning. In A. Ortony (Ed.), *Metaphor and thought* (pp. 438–461). Cambridge, England: Cambridge University Press.

Pintrich, P. R., Marx, R. W., & Boyle, R. A. (1993). Beyond cold conceptual change: The role of motivational beliefs and classroom contextual factors in the process of conceptual change. *Review of Educational Research, 63,* 167–199.

Posner, G. J., Strike, K. A., Hewson, P. W., & Gertzog, W. A. (1982). Accommodation of a scientific conception: Toward a theory of conceptual change. *Science Education, 66,* 211–227.

Roth, W. M. (1995). *Authentic school science.* Dordrecht, The Netherlands: Kluwer.

Rowell, J. A., & Dawson, C. J. (1985). Equilibration, conflict and instruction: A new class-oriented perspective. *European Journal of Science Education, 7,* 331–334.

Rubin, A. (1990). *Concreteness in the abstract: Dimensions of multiple representations in data analysis. Design for learning.* Cupertino, CA: External Research, Apple Computer.

Saxe, G. B. (1995). From the field to the classroom: Studies in mathematics understanding. In L. Steffe & J. Gale (Eds.), *Constructivism in education* (pp. 287–311), Hillsdale, NJ: Erlbaum.

Scott, P. H., Asoko, H. M., & Driver, R. H. (1992). Teaching for conceptual change:

A review of strategies. In R. Duit, F. Goldberg, & H. Niedderer (Eds.), *Research in physics learning: Theoretical issues and empirical studies* (pp. 310–329). Kiel, Germany: Institute for Science Education (IPN), University of Kiel.

Skemp, R. (1978). Relational understanding and instrumental understanding. *Arithmetic Teacher, 16,* 9–15.

Stavy, R. (1991). Using analogy to overcome misconceptions about conservation of matter. *Journal of Research in Science Teaching, 28,* 305–313.

Steffe, L. (1994). Children's multiplying and dividing schema. In G. Harel & J. Confrey (Eds.), *The development of multiplicative reasoning in the learning of mathematics* (590–592). Albany: State University of New York Press.

Steffe, L., von Glasersfeld, E., Richards, E., & Cobb, P. (1983). *Children's counting types: Philosophy, theory and application.* New York: Praeger.

Stepans, J., Dyche, S., & Beiswenger, R. (1988). The effects of two instructional models in bringing about a conceptual change in the understanding of science concepts by prospective elementary teachers. *Science Education, 72,* 185–195.

Stinner, A. (1992). Science textbooks and science teaching: From logic to evidence. *Science Education, 76,* 1–16.

Strike, K., & Posner, G. (1992). A revisionist theory of conceptual change. In R. Duschl & R. Hamilton (Eds.), *Philosophy of science, cognitive psychology, and educational theory and practice* (pp. 147–176). Albany: State University of New York Press.

Strube, P. (1989). The notion of style in physics textbooks. *Journal of Research in Science Teaching, 26,* 291–299.

Sutton, C. (1989). Writing and reading science: The hidden message. In R. Millar (Ed.), *Doing science: Images of science in science education* (pp. 137–159). London, England: Falmer Press.

Thompson, P. (1994). The development of the concept of speed and its relationship to concepts of rate. In G. Harel & J. Confrey (Eds.), *The development of multiplicative reasoning in the learning of mathematics.* Albany: State University of New York Press.

Toulmin, S. (1972). *Human understanding.* Princeton, NJ: Princeton University Press.

Voigt, J. (1985). Patterns and routines in classroom interactions. *Recherches en Didactique des Mathematiques, 6*(1), 69–118.

Vosniadou, S. (Ed.). (1994). Conceptual change in the physical sciences [Special issue]. *Learning and Instruction, 3,* 3–110.

White, R. T. (1986). Observations of a minor participant. In J. Baird & I. Mitchell (Eds.), *Improving the quality of teaching and learning—An Australian case study* (pp. 280–285). Melbourne, Australia: Monash University.

Yore, D. L. (1991) Secondary science teachers' attitudes towards and beliefs about science reading and science textbooks. *Journal of Research in Science Teaching, 28,* 55–72.

CHAPTER 8

Curriculum Development as Research: A Constructivist Approach to Science Curriculum Development and Teaching

Rosalind Driver and Philip H. Scott
University of Leeds, England

This chapter gives an account of a curriculum development initiative undertaken by the Children's Learning in Science (CLIS) Research Group. The project was carried out in collaboration with practicing science teachers over a period of three years, during which time teaching materials based on constructivist principles were devised for three topics (energy, particle theory, and plant nutrition), and were tried out and evaluated in classrooms. Teaching schemes for use in secondary schools were published for the three selected topics (CLIS, 1987) and a number of features that characterize constructivist approaches to teaching and learning were identified and documented (Driver, 1989a; Needham, 1987; Scott, 1987).

The chapter first outlines the theoretical and practical contexts in which this work took place and gives an overview of the project. This is followed by a more detailed account of one of the teaching schemes in order to illustrate how constructivist principles were implemented in the context of teaching the particulate theory of matter. The chapter concludes with implications of constructivist approaches for the development of science teaching schemes.

THEORETICAL AND PRACTICAL CONTEXTS

The constructivist perspective from which this project was undertaken (Driver & Oldham, 1986) asserts that individuals construct and restructure their schemes of the world, through their own mental activity, as a result of experience with phenomena and social interactions. Moreover, the informal knowledge schemes that students construct are not idiosyncratic. Research in a range of domains suggests that commonalities exist in these ideas so that it is feasible and important to consider ways of taking them into account in preparing curriculum materials. A further important aspect of the perspective argues that scientific knowledge itself is constructed socially. As a consequence, it is inappropriate to consider that students learn science through empirical inquiry simply by reading the "book of nature." Learning science involves being introduced to the "ways of seeing" of the scientific community and thus involves a socialization process (Driver, 1989b).

Learning science in school settings therefore is seen as a process that can involve replacement of informal ideas by a science view (conceptual change), but also might involve the development of scientific ideas alongside more informal viewpoints. In either case, making links between new (science) and old (informal) ways of looking becomes a crucial part of "sense making" for the learner, as does recognizing the context in which it is appropriate to use each. A further demand for the learner is one of coming to understand the different "rules" that govern each type of knowledge (Garrison & Bentley, 1990); for example, whereas science theories are judged (by the science community) in terms of their power to explain disparate phenomena, informal ideas often are accepted (by the lay public) on pragmatic grounds relating to specific contexts.

Our tasks in the project, thus, were to draw on the knowledge that exists about children's conceptions in the domains in question, to develop teaching approaches for promoting conceptual change and development consistent with a constructivist perspective, and to field-test these in normal science classrooms.

The project was undertaken as a form of action research in collaboration with secondary school science teachers. This practical context enabled us to draw on teachers' pedagogical expertise in devising teaching approaches, to provide facilities for field-testing approaches in classrooms with teachers who understood the rationale for the particular teaching approaches used, and to give credibility to the outcomes of the project among other teachers. The project was also part of the Secondary Science Curriculum Review in the United Kingdom, a four-year program that involved science teachers at local, regional, and national levels in a process of reviewing their practice. It was in this spirit of "reflection on practice" that science teachers were invited to take part in the CLIS project.

OUTLINE OF THE PROJECT PROGRAM

The program of work extended over three years and science teachers within traveling distance of Leeds University were invited to take part for at least a two-year period, during which time they would commit themselves to meetings out of school time and to try out materials in their schools (with children in the 13–15 age range). A group consisting of 10–12 teachers and a researcher was set up for each of the three topic areas. The topic areas of energy, particle theory, and plant nutrition were selected because they focus on concepts that have proved to be difficult for students and yet are central to an understanding of large areas of the sciences. Each of the three working groups followed the same outline program, which consisted of preparatory, development, and field-testing phases.

Preparatory Phase

During the first year, the researcher with each working group visited the classrooms of a number of the participating teachers when they were teaching the topic of interest and documented in some detail the existing curriculum provision and teaching approaches, including students' responses to particular learning activities. In addition, all the teachers collected information about their students' understanding of the topic using a diagnostic test. The information from the researcher and teachers was used to write case studies based on issues involved in teaching each of the topics (Bell, 1985; Brook & Driver, 1986; Wightman, 1986). These case study documents were reviewed by each working group toward the end of the year in order to diagnose the main conceptual difficulties students had in learning the topic and to critique the teaching approaches used.

Development Phase

Toward the end of this first year, the working groups identified the common prior conceptions that students tended to hold and agreed on goal conceptions for each of the three teaching schemes. The next task was to devise teaching approaches to help children with these intellectual demands. In order to do this, some general implications for teaching were drawn from consideration of constructivist perspectives on learning at a number of levels.

First, there were issues relating to the general nature of the *learning environment* in the classroom and the implications of this for classroom organization. The guiding concern here was to provide a learning environment that would be supportive of knowledge construction processes. Student group work was seen to be central in the social organization of the class as it provided necessary opportunities for individuals to interact with each other and the teacher, to review ideas, and to support or criticize each other in an informal setting. It also was considered important to have groups report at appropriate times to the whole class and, in these situations, the teacher's role was to facilitate structured discussion and, where necessary, to relate students' experiences and ideas to the science viewpoint.

A conceptual development view of learning informed decisions about the *structure of the teaching sequences.* In general, sequences were designed to start by taking account of the students' ideas, then using that information to develop activities to help students restructure their ideas; in the end, these led to opportunities for students to review any resulting changes in their conceptions.

The most detailed level of consideration was that of the *learning activity.* Knowledge of students' prior conceptions informed the design of specific learning activities. In some cases, learning activities involved practical work in which students' current ideas were extended or challenged. In other cases, theoretical tasks required students to generate theories to account for agreed-on events, or evaluate alternative theories relating to the same phenomenon.

Experimental teaching schemes were written in each of the three topic areas to reflect each of these levels of consideration. Teachers in each working group undertook to try out one of the schemes during the next school year.

Field-Testing Phase

During field testing, information about the conceptual progress that students were making in response to the experimental teaching was collected from a number of sources. Pre- and postteaching diagnostic tests were given to the students, who also were asked to keep informal learning diaries. On a lesson-by-lesson basis, teachers kept diaries, which later were analyzed by the researcher. In addition, detailed case studies of the learning taking place in two classrooms were undertaken by a researcher for each of the three topics. These case studies involved following target groups of students through the lessons and interviewing them periodically (see Johnston & Driver, 1991; Oldham, Driver, & Holding, 1991).

When most of the teachers had taught the topic, workshops for teachers and researchers were held to consider the data collected. Students' responses to each activity were reviewed in detail and, as a result, some activities were modified and others were replaced. The critical feature in the field-testing process was whether specific learning activities encouraged the required conceptual development. All three of the revised schemes were tried out again in a number of classrooms, following the same procedures, prior to final revision and publication.

TEACHING THE PARTICULATE THEORY OF MATTER FROM A CONSTRUCTIVIST PERSPECTIVE

In this section, an overview of the teaching scheme relating to particle theory (CLIS, 1987) is presented. The overall strategy adopted in the scheme is to draw on students' existing ideas about the properties and structure of matter and, through a series of carefully designed activities, help them toward an understanding of the accepted particle view. The aims for the scheme, agreed to by the working group, were that students should develop an understanding of particle theory for solids, liquids, and gases in terms of particle arrangement, particle motion, and forces between particles. Students also should be able to use these ideas to explain simple phenomena and thereby come to appreciate the explanatory power of scientific theories. A detailed account of the scheme in action and the research process that informed its development is given in Johnston and Driver (1991).

The teaching scheme is organized into the six sections shown in Figure 8.1. Each of the sections in the figure is reviewed below in terms of the learning activities used, the rationale behind those activities, and the progress in student learning documented during the trials.

Orientation and Elicitation of Student Ideas

In the introduction to the topic, students work in pairs and are presented with a number of activities each of which is designed to prompt thought and discussion about a specific property of matter. For example, students are asked: to open an air freshener and explain what they notice; to compare the compressibility of air, water, and sand in sealed syringes and to discuss their explanations for what they observe; to consider why blocks of different materials, but of identical shape, should have different masses. After working on each of six activities, pairs of students combine to form groups of four and each group presents their ideas about one of the phenomena on a poster.

This first part of the teaching sequence serves a number of different

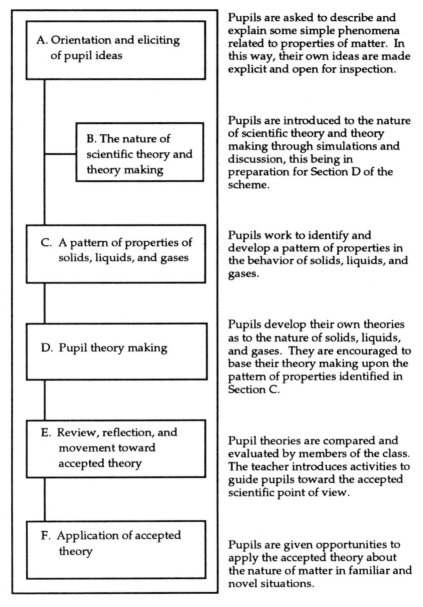

Figure 8.1. Outline of the Teaching Scheme for Particles

The content within the figure:

A. Orientation and eliciting of pupil ideas — Pupils are asked to describe and explain some simple phenomena related to properties of matter. In this way, their own ideas are made explicit and open for inspection.

B. The nature of scientific theory and theory making — Pupils are introduced to the nature of scientific theory and theory making through simulations and discussion, this being in preparation for Section D of the scheme.

C. A pattern of properties of solids, liquids, and gases — Pupils work to identify and develop a pattern of properties in the behavior of solids, liquids, and gases.

D. Pupil theory making — Pupils develop their own theories as to the nature of solids, liquids, and gases. They are encouraged to base their theory making upon the pattern of properties identified in Section C.

E. Review, reflection, and movement toward accepted theory — Pupil theories are compared and evaluated by members of the class. The teacher introduces activities to guide pupils toward the accepted scientific point of view.

F. Application of accepted theory — Pupils are given opportunities to apply the accepted theory about the nature of matter in familiar and novel situations.

functions. It provides shared experiences with phenomena, on which later theorizing about the structure of matter can be based. The activities were selected carefully to provide evidence for different aspects of the particle model. For example, the syringes prove to be very useful, and often are referred to by students in providing evidence for the relative spacing of particles in solids, liquids, and gases. Students indeed are intrigued by the "springiness" of air and the incompressibility of water. By offering a number of phenomena for students to consider, a broad range of evidence for later theory making and checking is created. The challenge for the students becomes one of developing a single theory to explain this disparate range of material properties.

Experience in many classrooms has shown that these initial activities have a powerful motivating influence on students. They become aware that it is a valid activity to examine closely, and to think about, common phenomena that often are taken for granted. They are asked for their points of view on these matters, and they become aware of their own thinking and of the fact that other students possibly are thinking in different ways. By such means, the context of the work begins to assume personal relevance and students are "drawn inside the problem." This is apparent from the following extract from a learning diary of one of the students who was 13 years old: "I liked the idea of thinking up my own ideas and sharing them with another person instead of being told about what something did and why, like a teacher lecturing to his class."

The posters presenting students' ideas are displayed prominently in the laboratory throughout the teaching so that those ideas are valued explicitly and are not lost from the "thinking agenda." At this stage, the teacher does not introduce the ideas of particles and few students spontaneously use particle ideas. One exception to this trend has been found in students' responses to the activity in which they consider why blocks of identical shape but of different materials have different masses. In this case, explanations based on particle packing have been quite common, as illustrated by the following student's comments: "We had to explain why block A weighed more than block B. I thought that it was something to do with atoms being compressed densely in A and not so packed in B." In this case, as in others like it, it would be wrong to assume that the student has a mature understanding of atomic theory. He obviously is aware of the existence of atoms and, for this particular phenomenon, offers an explanation in atomic terms. We do not know, however, whether the "atoms" referred to have the same nature and properties as those of the accepted science view, and caution is needed in interpreting such comments. It is noticeable that, at this stage, students can use particle ideas in response to this task but not for others.

The Nature of Scientific Theory and Theory-Making

This section invites students to consider through analogy the nature of scientific theories and theory-making. This is an important introduction to the later section in which students develop their own theories about the nature of matter. The work is based on two exercises, a simple "rule guessing" game and a group activity in which the students are given a series of "clues" to a murder mystery and are invited to find out "whodunit." Both of these exercises are carried out with little comment from the teacher and, on completion, the students are asked to reflect on how they were able to "spot the rule" and "solve the murder." Discussion leads to ideas of looking for patterns in data, the development of theories that satisfactorily account for available evidence, and a consideration of the use of imagination in generating theories that go beyond what is represented in the data. In some classes, the provisional nature of theories also has been explored by considering what might happen if new, and possibly contradictory, evidence becomes available.

After these ideas have been reviewed in the context of the two exercises, links are then made to science theories and theory-making. The idea that science theories provide an explanatory framework for dealing with a range of evidence about phenomena is emphasized, as is the idea that science theories in some cases might be prompted by investigating patterns in data.

A Pattern of Properties of Solids, Liquids, and Gases

The purpose of this section is to organize the varied and apparently disparate information about materials and to establish a pattern in the properties of solids, liquids, and gases that can be used as a starting point for student theorizing about the structure of matter.

The students first are presented with a range of artifacts and materials and are asked to classify each as solid, liquid, or gas. Whereas most of the items are nonproblematic (water, iron block, acetate strip, the air in the room), others tend to lead to lively debate (powders, cloth, shaving foam, ice). Many students, in carrying out this exercise, are prompted to reconsider their ideas about classifying matter. An example of this is a student who came to recognize, by a process of elimination, that nonrigid materials such as "cloth" are solids.

Almost without exception, students succeed in correctly classifying the full range of items. The next step is much more demanding. The students are asked to list those properties of solids, liquids, and gases that enable correct identification. ("What is it about solids that allows you to recognize that they are solid? . . . What is it about liquids . . . about gases?") The students work in groups to generate their "characteristic properties," and the ideas from the whole class are collected and displayed. These data then are used as a basis for identifying patterns in the properties of solids, liquids, and gases, including such features as compressibility and rigidity. Experience has shown that this sequence of activities can be problematic for both students and teacher. One common source of difficulty for students is that of identifying characteristic properties (properties that are both generalizable within a group and exclusive to that group) for each state of matter.

At the heart of this problem lies the fact that these students do not classify matter by referring to the characteristic properties of mutually exclusive groups; rather, they work through a process of comparison with prototypical examples of each state. It is hardly surprising, therefore, that students experience difficulty in isolating characteristic properties; it is important that the teacher recognize the nature of the intellectual demand being made.

At this point, the following kinds of statements from students are common: "Solids are hard"; "We can walk through gases"; and "Liquids are wet." The teacher faces an interesting pedagogical challenge in trying to move from these untidy and frequently situation-specific student ideas to establishing an agreed-on pattern of properties that will inform subsequent theory-making. This challenge has been approached through class discussion and review. Initially, the students might be invited to examine critically the properties appearing on the display and to suggest whether any should be deleted ("all solids aren't hard . . . what about that cloth"). Further negotiation and shaping leads to a pattern that might include elements such as "solids and liquids cannot be compressed, gases can be compressed"; and "solids have a fixed shape, liquids take the shape of their container, gases spread." Some interesting teacher comments on this section were forthcoming:

> Imposing "my pattern" on their data was not particularly easy. I found myself defending "my position" whilst trying not to let it look like that. The tidy patterns that I wished to introduce frequently were questioned by consideration of "everyday" properties of solids, liquids, and gases.

Student Theory-Making

This is the central activity of the sequence. The students draw on all of the previous sections as they work in groups to develop a theory about the structure of solids, liquids, and gases that will allow them to explain the previously identified pattern of properties: "What might solids, liquids, and gases be like inside such that they give rise to this pattern of properties?" Each group spends time reviewing available evidence and developing a theory before recording their ideas on a poster, which is presented to the rest of the class.

At this stage, it is common for groups to introduce particle ideas. The notion that matter (and, in particular, solids and liquids) is composed of particles, and that the properties of matter can be explained in terms of those particles, does not appear to be problematic for many students of this age. In trials with many classes, we have had only one class in which particle ideas were not suggested spontaneously and adopted as a "useful theory": Students made comments such as "The solid contains 'bits' that are packed together. The liquid has these bits which are spread out in between air. The bits in gas are spread out even more." "Solids cannot be compressed because the molecules already are packed tightly together."

In some ways, such fundamental notions of atomicity might be viewed as offering an "anchor" (Clement et al., 1987) for the subsequent development of ideas. At the same time, and as would be anticipated, there are clear points of divergence between "student" and "science" particle theories. We have found from trials in many classrooms that similar features tend to arise in students' particle theories. Students tend to develop static rather than dynamic particle models; they often have problems considering what lies between the particles (frequently considering that this must be air); they tend to consider that solids are strong and hold their shape because the particles are packed tightly (cohesive forces are not considered); problems emerge because of difficulties of scale (the size of the particles and the spaces between them). However, placing students in the position of theory-makers serves a number of positive functions. It puts students at the intellectual "sharp end" of coordinating theory and evidence. It also enables alternative interpretations of the particle model (e.g., that there is air between the particles) to be revealed explicitly so that these can then be picked up in the teaching. One student wrote in her diary: "The lesson was quite hard because, when we had to find out what was in the middle of a solid, liquid, or gas, I thought about particles. If you say that the particles are spread out in a gas, then what would be in between them? I thought that it is a very hard question." Perhaps the most impressive aspect of this statement is that the student should be in a position in which she was able to reflect on this gap in her own understanding. Intellectual involvement of this quality has been observed in a number of classrooms as students take on the role of theory-makers.

Review, Reflection, and Movement Toward Accepted Theory

In this section, each group presents its theory to the rest of the class and opportunities are provided for questioning and clarification of ideas. This is followed with activities selected by the teacher in order to develop those aspects of students' theories that are problematic.

The teacher must consider the nature of the differences between students' and accepted science theories and decide on appropriate activities to help the students toward the science viewpoint. The form of these activities varies with the nature of the issue being addressed. In the case of "what lies between the particles," the students are encouraged to think about, and to discuss, the implications of suggesting that there is air between the particles. Through such activity, students have been able to argue quite clearly that "there must be nothing between the particles; if there was air, then there would be particles of air filling the gaps." Here the teacher cannot offer empirical "proof" to solve the problem; the students must address the issue in terms of probing the logic and internal consistency of the emerging theory. At other times, the teacher will be able to, and will need to, introduce further evidence to prompt theory development. For example, if the student models are static representations, then the teacher can demonstrate diffusion effects in liquids and gases and encourage the students to consider the implications of those phenomena for their models.

Experience in a range of classrooms has shown that there are commonalities in the "conceptual pathways" (Scott, 1991; Scott, Asoko, & Driver, 1991) students follow in coming to terms with accepted particle theory. The conceptual pathways can be thought of as the "learning routes" along which students typically pass in developing understandings of a science domain. For example, many students move from believing that there is air between the particles to recognizing that there must be nothing. This pathway involves a change in thinking and is likely to be influenced by students' conceptions of the scale of the particle model. Many students have a fundamental notion of "air filling all of space" and struggle to relate that to a model in which there are "large empty spaces between air particles." Their problem is one of recognizing that the spaces between particles, at the same time, are relatively large but on a real-world scale are unimaginably small. In the case of particle motion, the pathways involve the introduction of the extra kinetic feature to the model. Students often have difficulties with this feature and question "what keeps the particles moving." Insights such as these into the nature of students' conceptual pathways have helped teachers in anticipating student difficulties and selecting activities that more closely support learning. Activities that address the most commonly occurring problems with the students' particle theories have been developed and are included in the scheme for use by teachers as appropriate.

Application of Accepted Particle Theory

In this final section of the teaching sequence, students review the various aspects of the particle theory that have been established, and the teacher then provides activities that allow them to apply those ideas. Some of these activities are similar to the initial phenomena, and students can compare differences in the nature of their explanations as they revisit tasks. Other activities require students to use particle ideas in new contexts. This phase of work aims to help students further develop their understanding of particle theory, to gain confidence in using it, and to recognize that their thinking has progressed as a result of teaching.

EVALUATION OF THE SCHEME OF WORK

Students' and teachers' reactions to the teaching scheme were documented. In general, teachers commented that teaching in this way involved more thought on their part, although they found it much easier and less stressful the second time through when they could anticipate both the kinds of ideas and the form of reasoning that tended to be used by students.

The extent to which the teaching had produced changes in students' conceptualizations was assessed by comparing students' responses to diagnostic questions set before and after teaching in two experimental classes. The results showed that there was an increase for over 50 percent of students in their use of particulate ideas in responding to the questions, although there was a smaller increase in the proportion of students using various features of the model. Details of these outcomes can be found in Johnston and Driver (1991). In general, students' reactions to the teaching approach have been interesting and encouraging.

CONCLUSION

The project has important implications for both curriculum development and teacher development. An important issue is the fundamental importance, from a constructivist perspective, of the field-testing process in curriculum development. In the past, materials that have been well thought through from the point of view of the scientific ideas they present and that are presented clearly to students might have been considered "good" curriculum materials. From a constructivist perspective, however, we cannot assume that what is taught is what is learned. Curriculum development and evaluation from this perspective must consider the understandings that learners are gaining from the activities. The process of iterative field testing, feedback, and revision was undertaken in developing materials that not only were teachable, but for which there was evidence of student progression in learning.

These issues point to curriculum development as being a legitimate subject for research. Such research would involve cycles of planning, implementing, and evaluating teaching activities and approaches with a view to maximizing the learning achieved. Although there were certain general strategies we found to be effective across the three domains, the actual learning activities are domain-specific and hence any research-based curriculum development has to be undertaken domain by domain across the science curriculum.

With regard to teacher development, the project prepared teachers to use a broader range of teaching strategies and activities. Although group work was a new approach for many teachers and carried its own set of attendant worries, by the end of the project, the widely held belief among the teachers was that group work is successful in engaging students and promoting learning. Regular teacher group meetings provided an important source of mutual support and encouragement, as well as the opportunity to discuss just how students had responded to activities.

Teachers responded positively to various teaching strategies and activities in the context of "putting into action" a constructivist perspective on learning. Many of the teachers found intellectual and professional satisfaction in having some kind of theoretical framework to help guide their actions. Indeed, it is not irrelevant to note that quite a number of the teachers who were involved in the project continue to undertake research work in their own classrooms. One interesting feature of this work has been the way in which the processes of reflection and analysis have continued long after the initial enthusiasms of the project have subsided. Very recently, a group of Leeds teachers and researchers discussed at length what precisely is "going on" in a certain part of the particles scheme and made progress in their own thinking in doing so. Teaching, reflection, and learning continue—and it is not just students who are learning.

ACKNOWLEDGMENT

The research reported here was supported by a grant from the Department of Education and Science, 1985–1987.

REFERENCES

Bell, B. (1985). *The construction of meaning and conceptual change in classroom settings: Case studies on plant nutrition.* Leeds, England: Children's Learning in Science Project, Centre for Studies in Science and Mathematics Education, University of Leeds.

Brook, A., & Driver, R. (1986). *The construction of meaning and conceptual change in classroom settings: Case studies on energy.* Leeds, England: Children's Learning in Science Project, Centre for Studies in Science and Mathematics Education, University of Leeds.

Clement, J., et al. (1987). Overcoming students' misconceptions in physics: The role of anchoring intuitions and analogical validity. In J. Novak (Ed.), *Proceedings of the Second International Seminar on Misconceptions and Educational Strategies in Science and Mathematics* (Vol. 3, pp. 84–94). Ithaca, NY: Cornell University.

Children's Learning in Science (CLIS). (1987). *CLIS in the classroom: Approaches to teaching.* Leeds, England: Centre for Studies in Science and Mathematics Education, University of Leeds.

Driver, R. (1989a). Changing conceptions. In P. Adey (Ed.), *Adolescent development and school science* (pp. 79–103). London, England: Falmer Press.

Driver, R. (1989b). The construction of scientific knowledge in school classrooms. In R. Millar (Ed.), *Doing science: Images of science in science education* (pp. 83–106). London, England: Falmer Press.

Driver, R., & Oldham, V. (1986). A constructivist approach to curriculum development in science. *Studies in Science Education, 13,* 105–122.

Garrison, J. W., & Bentley, M. L. (1990). Science education, conceptual change and breaking with everyday experience. *Studies in Philosophy and Education, 10,* 19–35.

Johnston, K., & Driver, R. (1991). *A case study of teaching and learning about particle theory: A constructivist scheme in action.* Leeds, England: Children's Learning in Science Research Group, Centre for Studies in Science and Mathematics Education, University of Leeds.

Needham, R. (1987). *Teaching strategies for developing understanding in science.* Leeds, England: Children's Learning in Science Project, Centre for Studies in Science and Mathematics Education, University of Leeds.

Oldham, V., Driver, R., & Holding, B. (1991). *A case study of teaching and learning about plant nutrition: A constructivist scheme in action.* Leeds, England: Children's Learning in Science Research Group, Centre for Studies in Science and Mathematics Education, University of Leeds.

Scott, P. H. (1987). *A constructivist view of learning and teaching in science.* Leeds, England: Children's Learning in Science Project, Centre for Studies in Science and Mathematics Education, University of Leeds.

Scott, P. H. (1991). Conceptual pathways in learning science: A case study of the development of one student's ideas relating to the structure of matter. In R. Duit, F. Goldberg, & H. Niedderer (Eds.), *Research in physics learning: Theoretical issues and empirical studies* (pp. 203–224). Kiel, Germany: Institute for Science Education (IPN), University of Kiel.

Scott, P. H., Asoko, H., & Driver, R. (1991). Teaching for conceptual change: A review of strategies. In R. Duit, F. Goldberg, & H. Niedderer (Eds.), *Research in physics learning: Theoretical issues and empirical studies* (pp. 310–329). Kiel, Germany: Institute for Science Education (IPN), University of Kiel.

Wightman, T. (1986). *The construction of meaning and conceptual change in classroom settings: Case studies on the particulate nature of matter.* Leeds, England: Children's Learning in Science Project, Centre for Studies in Science and Mathematics Education, University of Leeds.

CHAPTER 9

Strategies for Remediating Learning Difficulties in Chemistry

Ruth Ben-Zvi and Avi Hofstein
Weizmann Institute of Science, Rehovot, Israel

The inadequacy of students' knowledge of science and the resulting learning difficulties have been a central theme of research in the last decade. The initial phases focused on the description of students' difficulties, with numerous studies in the area of concept learning showing that students frequently hold ideas that are different from the accepted scientific view (Driver, 1988; Duit, Jung, & von Rhoneck, 1985; McDermott, 1984; Tobin, 1993). In some cases, these ideas are consistent and make sense from the student's point of view; in other cases, students demonstrate a confused, inconsistent way of thinking.

So far, however, research findings in this area have had only limited influence and impact on the practice of science education in the classroom or school laboratory, let alone on the design of curricula and teaching programs. There are several reasons for this state of affairs (Kempa, Ben-Zvi, Hofstein, & Cohen, 1988). First, researchers themselves frequently have failed to make explicit the implications of their research findings for educational practice and curriculum design. Second, investigations into students' learning difficulties often are conducted under conditions that are remote from the reality of teaching. Although findings from such investigations are usually interesting in themselves, the artificiality of the setting in which they were obtained can represent a barrier to their practical application. Third, educational practitioners (including teachers and teacher educators, as well as curriculum developers) tend to be reluctant to accept and act on the basis of research findings, either because they conflict with personal beliefs and convictions or because the changes they call for are difficult to implement in practice. Fourth, teachers frequently are unaware of the fact that learning difficulties and alternative conceptions exist among their students.

The purpose of this chapter is to describe an approach to the design, development, and implementation of a new curriculum in chemistry that takes into account students' learning difficulties and includes strategies for remediating those learning difficulties. Two sources of information, namely, the results of matriculation examinations and the diagnosis of student learning in tenth-grade chemistry, are presented to illustrate how insights were obtained about students' learning difficulties using the concept of structure and bonding as an example.

DEFINITION OF LEARNING DIFFICULTIES

Learning difficulties can be said to exist in any situation in which a student fails to grasp a concept or idea as the result of one or more of three factors. The first factor concerns the nature of the ideas/knowledge system already possessed by the student, or the inadequacy of such knowledge in relation to the concept to be learned (Hewson & Hewson, 1984). Indeed, according to Ausubel, Novak, and Hanesian (1978), one of the most important single factors influencing learning is what the learner already knows. Similarly, Head (1985) claims that individuals' prior conceptions derive from experience with the environment and their existing ideas, which are used to model new situations. The second factor concerns the demand and complexity of a learning task in terms of the information-processing requirements compared with the student's information-handling capacity. The third factor concerns communication problems arising from language use, especially in relation to technical terms, general terms with context-specific specialized meanings, and the complexity of the sentence structure and syntax used by the teacher, compared with the student's own language.

In order to bring the work on learning difficulties to the stage at which it is useful in practice, several steps are necessary (Eylon, 1988): diagnosis of learning difficulties and how they are manifested in student behavior; interpretation of the difficulties with a focus on their source; treatment of the difficulties by introducing teaching approaches that lead to avoidance, by-passing, or remediation of a difficulty; and evaluation aimed at assessing the educational effectiveness of the teaching approach. It is suggested that the implementation of one's knowledge concerning a learning difficulty in a curricular context proceeds in a cycle as suggested in Figure 9.1.

To accomplish the steps in Figure 9.1, there is a need to build coopera-

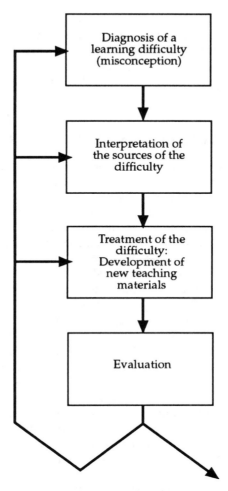

Figure 9.1. Curricular Interpretation of Learning Difficulties

tion between the four dominant groups involved in curriculum development and implementation: researchers, curriculum developers, decision makers (i.e., inspectors or supervisors who are responsible for implementing syllabi), and science teachers in school. This model was adapted and used in the development of materials by a curriculum development team at the Weizmann Institute of Science as a part of a science curriculum development network located at the Amos de Shalit—Israel Science Teaching Center.

In Israel, the academic senior high school consists of three years (age range is 15–18 years). Chemistry is a compulsory subject in the first year (grade 10) and taking advanced courses is optional. Those who opt for advanced courses in chemistry can be examined in the matriculation examinations, taken at the end of grade 12, at either the moderate or the high level.

USE OF MATRICULATION EXAMINATION RESULTS

Important sources of information for curriculum change in Israel are the results and analyses of chemistry matriculation examinations (Bagrut). Milner (1983) showed that one of the sources for many learning difficulties is the textbook, particularly its content and structure, as well as the teaching and learning methods employed. The conclusions from the analysis of students' answers in matriculation examinations can be used in various ways (e.g., in the development of the new curriculum as described in the next section, at teachers' conferences, during in-service training, and when rewriting various teachers' guides). Two examples of questions taken from the Bagrut examinations and aimed at assessing students' understanding of the concept of atomic and/or molecular structures are shown in Figure 9.2.

In Example 1, 63 percent of the students gave wrong answers. Suggested reasons for such a high percentage of wrong answers are, first, unclear differentiation between inter- and intramolecular bonds and, second, that students did not understand that the term "bond energy" is defined as the amount of energy needed to break a chemical bond. In other words, students presented a vague understanding of the interrelationship between the concept of energy and the concept of bonding.

In Example 2, students who chose distractors (a) and (b) did not understand the formulas as representing substances that consist of many particles. Students who chose distractor (d) did not understand that O_2 represents a molecule of oxygen and that CO_2 means two atoms of oxygen bound to an atom of carbon. Similar misconceptions were identified by Peterson, Treagust, and Garnett (1989) with grade 11 and 12 Australian high school students.

DIAGNOSIS OF STUDENT LEARNING DIFFICULTIES IN INTRODUCTORY CHEMISTRY

This section follows the stages presented in Figure 9.1 in describing a diagnostic study of chemistry students' learning difficulties in introductory chemistry.

Example 1

C_2H_6O is the molecular formula of two isomers: ethanol (CH_3CH_2OH) and dimethyl ether (CH_3OCH_3). The heat of combustion of the alcohol is lower than that of the ether. What is the reason for this phenomenon?

	% of students choosing
a) In ether, there are high energy bonds.	40
b) In the alcohol, more energy is used to form the bonds.	12
c) The alcohol has a higher boiling temperature.	11
d) In the molecule of alcohol, stronger bonds exist between atoms. (correct answer)	37

Example 2

The following information relates to substances in room temperature and pressure. Which of the following is correct?

	% of students choosing
a) Solid iodine ($I_{2(s)}$) consists of one molecule, made up of two iodine atoms, attached to each other by a chemical.	16
b) Solid sodium ($Na_{(s)}$) consists of an atom that is not bound to another atom.	6
c) Ammonia gas ($NH_{3(g)}$) consists of many molecules in which each of the hydrogen atoms is bound to an atom of nitrogen. (correct answer)	71
d) Carbon dioxide ($CO_{2(g)}$) consists of an atom carbon bound to a molecule of oxygen ($O_{2(g)}$).	7

Figure 9.2. Examination Questions Assessing Students' Understanding of Atomic and/or Molecular Structures

Diagnosis of a Learning Difficulty

A sample of 337 students from 11 tenth-grade classes that studied chemistry for a period of eight months were given a set of formulas (e.g., O_2, $O_{2(g)}$, N_2O_4, NO_2) and asked to describe these by drawing models. When students' views about structure were examined, most of the students, who were at an advanced stage of the introductory course in chemistry, knew how to represent one molecule of an element. However, many of the students had difficulty representing correctly one molecule of a compound or an element in the gaseous or solid state. A third of the sample represented incorrectly the structure of a molecule N_2O_4 (e.g., as two connected disconnected fragments—one denoting N_2 and the other denoting O_4). About a third of the sample represented incorrectly an element in the gaseous state (e.g., in the description of $O_{3(g)}$, most of the wrong answers represented the gas by one molecule or by three disconnected atoms). The performance dropped even more when students were asked to represent a compound in the gaseous state (70 percent incorrect).

Interpretation of the Sources of the Difficulty

The results presented here, which are part of a more comprehensive study, led us to assume four main reasons for students' alternative conceptions about structure. The first reason is nonscientific understanding of the atomic model. If students think that the atom is a small piece of an element, then an additive view of the structure of a compound (i.e., one small piece of one element being near a small piece of another element) is a natural outcome. The second reason is a misleading use of models, which in textbooks is usually confined to single units. For example, the representation of a chemical reaction, such as synthesis of HCl, is presented by one molecule of hydrogen, one molecule of chlorine, and two molecules of hydrogen chloride. Third, students misunderstand chemical equations, reading them as representing single units and not moles of units. The fourth reason for alternative conceptions is information overload. As described above, when students had to represent a compound in the gaseous state, their performance dropped noticeably. In order to perform correctly, students had to coordinate two aspects, each of which caused difficulties by itself: the transition from an element to a compound, and transition from one molecule to many molecules. The demands of the task apparently overloaded students' working memory and they regressed to simpler, incorrect models by neglecting one aspect or another (Eylon, Ben-Zvi, & Silberstein, 1986).

Treatment of the Difficulty: Development of New Teaching Materials

In our attempt to overcome the difficulties mentioned in the previous section, an analysis of students' answers concerning the atomic model led us to assume that the intuitive models that students hold (i.e., that an atom is a piece of substance carrying all of the properties of this substance) placed them in the ancient Greek period. Because it was felt that many of the other difficulties were caused by this view, the objective of the developers of the program was to demonstrate to students how and why the atomic model was changed. The atom therefore is presented as an ever-developing model, as shown in Figure 9.3, whose characteristics change in accordance with new facts that have to be explained.

Following an introduction to scientific theories and models, a brief historical review is presented leading to a model of the atom that was devised in the beginning of the nineteenth century in order to explain some quantitative aspects of chemistry known at the time. This model, the Dalton Atom, could explain the laws of constant composition and multiple proportions but could not explain other properties of matter such as electrical conductivity. Dalton's ideas, therefore, had to be changed to include additional features. This interplay of facts/model/ new facts/new model is carried on during some months of study and with a summary of the stages of the development of the atomic model.

In the new textbook, much stress is put on presenting many particles whenever possible (using models of solids, liquids, and gases). Whenever the representation of many particles tends to obscure the point under discussion, the student is told specifically that, for reasons of clarity, only one set of particles is presented. Almost every chapter is followed by a unit called "the chemists' language," in which the meaning of symbols and equations is taught. Here again, the need to think simultaneously of many particles is stressed. Another factor that was found to affect students' performance was information overload. It is our belief that, if students can get used to thinking of many units and of the structure of each unit, then they have fewer problems concerning a scientific understanding of aspects of structure. This treatment is likely to help reduce the overload on students' short-term memory.

Evaluation of the New Curriculum

An intensive evaluation was conducted with 1,078 students from 35 classes to find out about the educational effectiveness of the new teaching materials. Half of the sample ($n = 538$) studied the currently used book (control group), and the remainder ($n = 540$) (experimental group) stud-

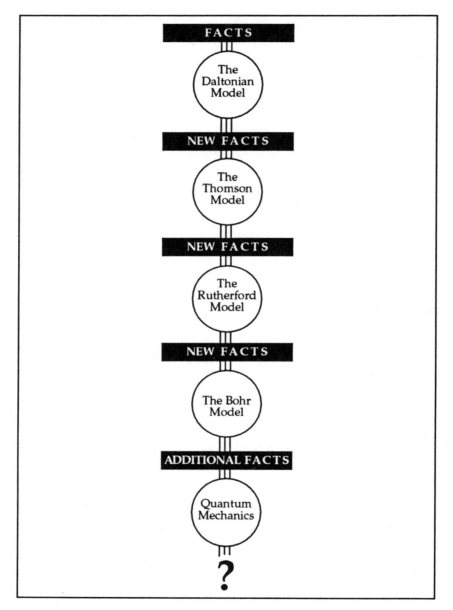

Figure 9.3. Development of the Atomic Model

ied the new course, Chemistry—A Challenge (Ben-Zvi & Silberstein, 1984). The students were given three sets of questionnaires, each consisting of achievement tests and diagnostic questions to which they had to respond by drawing models. One set was given prior to the beginning of their study and tested students' understanding and knowledge of their previous studies. The second set was given in the middle of the year and the third set served as a posttest. By the end of the year, the high achievers succeeded in the tasks no matter which learning method was used. The new program, however, had a pronounced effect on the middle and lower achievers, who scored higher on the tasks than did those students who studied from the currently used book.

CONCLUSION

In this chapter, an attempt was made to narrow the gap that exists between research on students' learning difficulties and misconceptions and the classroom practice of science education in general and education in chemistry in particular. The potential benefit for both teachers and curriculum developers in exploring links between research and practice in the area of student learning difficulties is that it could lead to the formulation of rules or guidelines that are directly applicable in the design of instruction in science curriculum development.

Research suggests that learning difficulties in chemistry are caused by: deficiencies/inadequacies in students' knowledge structure (e.g., Ben-Zvi, Eylon, & Silberstein, 1987; Kempa & Hodgson, 1976); the demand and complexity of learning tasks in terms of information processing compared with students' information-handling capacity (e.g., Eylon, 1988; Hewson & Hewson, 1984); the use of analogies and scientific language and related communication problems in science education (e.g., Maskill, 1988; Thiele & Treagust, 1994); and information overload (e.g., Johnstone & El-Banna, 1986). No single aspect can account adequately for the whole spectrum of learning difficulties and their underlying causes. Some can be explained by the constructivist theory of learning and others by the information-processing model—together they provide a theoretical basis for the interpretation and analysis of memory- and language-related learning difficulties. With this knowledge in mind, we must extend our inquiries from the mere identification and recording of learning difficulties to the generation and evaluation of strategies and curriculum materials that can be used to overcome or at least reduce such difficulties.

ACKNOWLEDGMENTS

Much of this study was conducted in collaboration with J. Silberstein and B. Eylon of the Department of Science Teaching, Weizmann Institute of Science. The authors are indebted to these two professors for their help.

REFERENCES

Ausubel, D. P., Novak, J. D., & Hanesian, H. (1978). *Educational psychology: A cognitive view.* New York: Holt, Rinehart, and Winston.

Ben-Zvi, R., Eylon, B., & Silberstein, J. (1987). Students' visualization of a chemical reaction. *Education in Chemistry, 24,* 117–120.

Ben-Zvi, R., & Silberstein, J. (1984). *Chemistry: A challenge.* Rehovot, Israel: Department of Science Teaching, Weizmann Institute of Science.

Driver, R. (1988). Theory into practice II: A constructivist approach to curriculum development. In P. Fensham (Ed.), *Development and dilemmas in science education* (pp. 133–149). London, England: Falmer Press

Duit, R., Jung, W., & von Rhoneck, C. (1985). *Aspects of understanding electricity: Proceedings of an international workshop.* Kiel, Germany: Institute for Science Education, University of Kiel.

Eylon, B. (1988). Towards a typology of learning difficulties. In R. F. Kempa, R. Ben-Zvi, A. Hofstein, & I. Cohen (Eds.), *Learning difficulties in chemistry* (Proceedings of a Bi-National UK–Israel Seminar) (pp. 33–53). Rehovot, Israel: Weizmann Institute of Science. (Obtainable from Avi Hofstein, Department of Science Teaching, Weizmann Institute of Science, Rehovot 76100, Israel)

Eylon, B., Ben-Zvi, R., & Silberstein, J. (1986, March). *Active hierarchical organization: A vehicle for promoting recall and problem solving in introductory chemistry.* Paper presented at the annual meeting of the National Association for Research in Science Teaching, San Francisco, CA.

Head, J. (1985). *The personal response to science.* Cambridge, England: Cambridge University Press.

Hewson, P. H., & Hewson, M. G. (1984). The role of conceptual conflict in conceptual change and the design of science instruction. *Instructional Science, 13,* 1–13.

Johnstone, A. H., & El-Banna, H. (1986). Capacities, demands, and processes: A predictive model for science education. *Education in Chemistry, 23,* 80–86.

Kempa, R. F., Ben-Zvi, R., Hofstein, A. J., & Cohen, I. (1988). *Learning difficulties in chemistry* (Proceedings of a Bi-National UK–Israel Seminar). Rehovot, Israel: Weizmann Institute of Science.

Kempa, R. F., & Hodgson, G. H. (1976). Levels of concept acquisition, and concept motivation in students' chemistry. *British Journal of Educational Psychology, 46,* 253–260.

Maskill, H. R. (1988). Science teaching and classroom language. In R. F. Kempa, R. Ben-Zvi, A. Hofstein, & I. Cohen (Eds.), *Learning difficulties in chemistry*

(Proceedings of a Bi-National UK–Israel Seminar) (pp. 165–167). Rehovot, Israel: Weizmann Institute of Science.

McDermott, L. C. (1984). Research on conceptual understanding in mechanics. *Physics Today, 37*, 24–32.

Milner, N. (1983). *Analysis of students' learning difficulties in chemistry as manifested in the matriculation examination.* Unpublished MSc dissertation, Department of Science Teaching, Weizmann Institute of Science (in Hebrew).

Peterson, R. F., Treagust, D. F., & Garnett, P. (1989). Development and application of a diagnostic instrument to evaluate grade 11 and 12 students' concept of covalent bonding and structure following a course of instruction. *Journal of Research in Science Teaching, 26*, 301–314.

Thiele, R. B., & Treagust, D. F. (1994). An interpretive examination of high school chemistry teachers' analogical explanations. *Journal of Research in Science Teaching, 31*, 227–242.

Tobin K. (1993, January). *Critical perspectives on constructivism, power, and mediation of science learning.* Paper presented at the conference entitled "Science Education in Developing Countries: From Theory to Practice," Jerusalem, Israel.

CHAPTER 10

Using Student Conceptions of Parallel Lines to Plan a Teaching Program

Helen Mansfield
University of Hawaii, Honolulu, United States

John Happs
Learning Performance Seminars, Perth, Australia

There is now a realization that teaching in mathematics and science generally has failed to overcome a number of cognitive barriers to student learning (Carpenter & Fennema, 1988; Kempa, 1993; Peterson, 1988). Despite several decades of innovation in curriculum content and design, there is still scant evidence to suggest that student understanding in these subjects has improved significantly. Traditional teaching strategies generally do not recognize students' conceptions and often fail to take into account the meaning of specific words as they are used and understood by classroom teachers and students. Recent approaches to teaching have proceeded by taking into consideration students' prior knowledge and how this is likely to mediate the learning process. If teaching strategies are to be seen as successful, they should lead to student understanding and retention of new knowledge over an extended period of time so that the classroom teacher does not have to reteach concepts.

The research described in this chapter probed and documented student conceptions about aspects of parallel lines. The information obtained was used to design, test, and refine materials and specific teaching strategies that drew on the ideas that students bring with them to the classroom. Twelve teachers in 10 different secondary schools taught and evaluated the final teaching program. Student understanding was assessed before, during, and after the teaching program, and delayed posttests were used to determine whether there was long-term retention or regression of ideas.

In order to develop more effective teaching strategies, we had to consider such factors as the central ideas that students bring with them to the learning situation. Evidence suggests that these ideas can override the knowledge being presented in class, distort new information in unanticipated ways, or coexist with new information (Osborne & Wittrock, 1983). In the last case, students simultaneously hold two views, one used in the mathematics classroom and the other in the outside world. Students can give a different meaning to a word than that intended by the teacher (Vinner & Dreyfus, 1989). If the student's meaning for a word is more limited or more extensive than the teacher's, or is quite idiosyncratic, then information given by the teacher can be misconstrued and used in nonmathematical ways.

If content understanding is evaluated immediately following instruction, it can indicate that students appear to have a new understanding because they have demonstrated an ability to pass a test using short-term recall. However, it does not necessarily mean that the students have accepted and assimilated this new understanding, and it does not indicate that individual students are able to draw on this new understanding in the longer term. In order for teaching strategies to be considered successful, they should bring about long-term conceptual change. For the purposes of this chapter, we refer to long-term as a period of at least six months following instruction. Our teaching strategies take into consideration student conceptions and personal meanings for words that mathematicians would be likely to use in a specific way in the context of parallel lines. Our teaching strategies had to challenge prevalent conceptions and give students an appreciation of any potential ambiguity over specific words. The phases of our research indicate the ways in which we took into consideration those problems likely to impede long-term conceptual change with respect to the topic of parallel lines.

PHASE 1: IDENTIFYING STUDENTS' CONCEPTIONS ABOUT PARALLEL LINES

Initially, we interviewed 37 eighth-grade students at two Western Australian secondary schools. In order to find out the meanings students had for words related to parallelism and to gain some insight into students' conceptions about parallel lines, we interviewed the students individually and asked them to give their meaning of the term "parallel" with examples. Students were asked to state whether each of 10 line drawings showed examples of parallel lines (Mansfield & Happs, 1987). Students used terminology such as "alongside," "next to each other," "not meeting," and "going on forever" to describe parallel lines. While the students often used similar everyday terminology to describe what they considered to be properties of parallel lines, the meanings of these terms as used by the students frequently were different.

Students' responses indicated that many thought of parallel lines in terms of equidistance rather than nonintersection. With this idea of parallel lines, some students thought that curved lines could be parallel provided that they are equidistant. Railway lines frequently were named as an example of parallel lines. Specifically, students thought that curved lines can be parallel; concentric circles can be parallel; converging line segments can be parallel, provided that their intersection is not shown on the page; segments can be parallel only if they are the same length and alongside one another; curves and straight lines can be parallel if they are equidistant at similar points, such as the turning points of a sine curve; intersecting curves can be parallel if they are equidistant at similar points; and parallel lines can be identified by global appearance rather than by accurate measurement (Mansfield & Happs, 1987). These students' conceptions contrasted with the mathematical view generally presented in school texts, which restricts the term *parallel* to straight lines, so that concentric circles and curves cannot be regarded as parallel. Similarly, converging line segments are not parallel, even when no point of intersection is present.

PHASE 2: DEVELOPMENT OF THE TEACHING PROGRAM

The teaching strategies used in the program were based on a constructivist view of learning, which holds as a principle that students construct mathematical ideas through reflection on their actions within a mathematical context (Cobb, 1988, 1990; Resnick, 1989; von Glasersfeld, 1987). These ideas were tested against the ideas of other students in the group and against the mathematical view presented by the teacher or the text. In addition, the students' ideas were tested against the constraints of the mathematical situation itself.

We designed a teaching program of four 50-minute lessons to encourage students to articulate their own ideas about parallel lines through discussion of the line drawings. Discussion was initiated by asking groups to sort 10 line drawings into categories of parallel, not parallel, and unsure. We later showed each group how a mathematician would categorize the line drawings. Because they were responsible for formulating group responses, the students had to consider the ideas that others held about parallel lines. At the same time, they were encouraged to defend their own ideas. In the case of students who did not appear to have a well-formulated concept of parallel lines, the group work prompted them to construct their own meaning of the word "parallel," perhaps for the first time.

Our initial teaching program was taught by the first author in one eighth-grade class. The second author interviewed three girls and three boys from one group before and after the teaching program, and monitored the interactions of this group with the teaching materials and the teacher. As a result of data collected from this trial, the teaching program was revised to include an investigation of coplanarity (i.e., being in the same plane) and the development of accurate measuring skills. The revised teaching program then was used by two eighth-grade teachers in a different secondary school to teach the topic of parallel lines. A group of three girls and three boys in each class was interviewed before and after the teaching program, and monitored throughout by the authors. The teaching materials and plans for four lessons were provided for the teachers. The materials included the relevant overhead transparencies and worksheets, as well as resources such as shapes for tessellations (mosaic patterns). A brief written rationale was provided for each lesson, and the lesson plans outlined the procedures to be followed but were not scripted. This allowed the teachers to use their own preferred explanations, styles of teaching, and classroom management while they used the strategies that had been developed.

Throughout the program, students worked in groups, which facilitated discussion and was useful to us in monitoring the interactions of a group within the class. The teaching strategies that we used led to different learning outcomes for the students whom we monitored. We were successful in increasing the number of correct responses to the items that we used; for example, students who thought initially that curves might be parallel became convinced that they are not. Nevertheless, some real-life situations such as railway lines still posed difficulties because students previously had heard teachers use them as examples of parallel lines. We may have succeeded in giving some students a fairly clear view of parallel lines for use in school, while leaving intact their earlier view for use out of school. We suspect that these students could operate with separate conceptual frameworks to the extent that they use a mathematical perspective in the classroom and their personal construction of reality outside.

While we were successful in convincing students that parallel lines are straight, we found that there still existed some conceptual and mechanical barriers to construction of the accepted mathematical view. Some of the students still were unsure of the parallelism of nonaligned segments, and the distinction between lines and segments caused some difficulty. A final interview with these students was carried out one month after the teaching program. In that month, the students maintained their modified understanding of the concept and, in two cases, enhanced it by recognition of the necessity for parallel lines to be coplanar.

For the two teachers who used our materials and teaching program, the requested classroom organization in which students worked in groups meant quite a substantial change in teaching style with which they were not entirely comfortable. Despite this, we decided not to change this aspect of the teaching program because we considered group work to be underpinning to our approach to conceptual change. The two teachers provided other valuable feedback on the program, and suggested further modifications, which were used to prepare the final teaching program.

PHASE 3: THE FINAL TEACHING PROGRAM

After the development, trial, and revision of the teaching program in Phase 2 of the research, we considered that the lesson plans and teaching materials were successful in challenging many student conceptions about parallel lines. The final lesson plans, materials, and rationales were given to mathematics teachers in 10 schools. Twelve eighth-grade classes in these schools then were taught the topic of parallel lines using our teaching program (Mansfield & Happs, 1992), which included the following steps:

1. Working in groups, the students sorted 10 items into groups of those that were parallel, those that were not parallel, and those about which they were unsure. The students' responses were recorded, and the teacher introduced the mathematical view of each item. Students compared their own view with the mathematical view. A class definition of parallel was constructed.
2. Students were introduced to the need for parallel lines to be coplanar by consideration of parallel lines and skew lines drawn as edges on a box.
3. Students were introduced to the use of file cards to measure the distances between lines. The file cards provided a right angle to ensure that measurement was perpendicular to the lines.

4. Students identified sets of parallel lines in tessellation patterns that they constructed.
5. Students identified ladder and zigzag patterns in pictures of real-life examples and in sets of parallel lines and discovered through comparison that corresponding angles and alternate angles are equal. The ladder-and-zigzag approach was drawn from the research undertaken by Fuys, Geddes, and Tischler (1984).
6. Students used corresponding and alternate angles to develop an argument that the angle sum of a triangle is 180 degrees.
7. Students used their knowledge of angle properties to solve problems involving parallel lines.
8. Students in groups summarized properties of parallel lines and of angle relationships found in sets of parallel lines.

The lessons were implemented following a written pretest on parallel lines that was administered to all students involved in the program. A posttest was administered to all students immediately following instruction. Both pretests and posttests consisted of a series of line drawings for students to categorize as examples or nonexamples of parallel lines, a series of true/false questions dealing with the properties of parallel lines, and an exercise on concept mapping in which students were asked to link given concepts with their own propositional statements about parallel lines (Mansfield & Happs, 1989, 1991). Results from these tests indicated that most students had altered their initial conceptions and constructed new conceptual frameworks that accommodated the common mathematical perspective. The teachers provided written feedback about their perceptions of the program, and made comments about the range of student conceptions identified and the ease with which the materials and lesson plans could be implemented within the normal constraints of the school setting.

PHASE 4: MONITORING LONG-TERM CONCEPTUAL CHANGE

Six months after using the program, students from one school were interviewed in order that long-term understanding and retention of ideas about parallel lines might be assessed (Happs & Mansfield, 1989). The results of this delayed posttest were encouraging in that most students demonstrated long-term modification of their initial conceptions about parallel lines with a perspective more closely aligned to the standard mathematical view.

Two years after the program, a group of students from a different

Table 10.1
Experimental and Control Group Means on
Three Student Tests on Parallel Lines

	Mean	
Test	Experimental Group	Control Group
Line drawings[a]	7.7	6.6
True/false questions[b]	10.4	10.4
Concept maps		
Correct propositions	5.5	3.4
Incorrect propositions	2.1	1.4
Concepts included	8.3	6.9

Note. The experimental group and the control group each contained 20 students.
[a]Number of line drawings in the test was 10.
[b]Number of true/false questions was 14.

school that had participated in the teaching program on parallel lines was matched with a control group that had not received instruction on parallel lines in the earlier phase of our study. The two groups were matched on the basis of gender and the mathematics units completed in the first three years of secondary schooling (Mansfield & Happs, 1991). The same written tests and concept map tests were used with all the students.

Although the experimental group had constructed concept maps in the pretest and posttest associated with the teaching program two years earlier, this was the first time the control group had constructed a concept map. Consequently, we acknowledge that the experimental group had greater familiarity with this task. Comparisons were made between the mean scores of the two groups in response to the three tests used in this phase of the study. The results, shown in Table 10.1, indicate that, although the experimental group performed better than the control group in their responses to the questions involving line drawings, the two groups performed equally well on the true/false questions. The experimental group obtained mean scores on the concept map test which were superior to the scores produced by the control group. By contrast, the experimental group produced more incorrect propositions than did the control group.

It is possible that the greater number of incorrect propositions given by the experimental group could have resulted from factors other than a greater lack of understanding about parallel lines. Because the experimental group was more experienced with the task of constructing concept maps, they might have had more test time to reflect on the task and include more propositions, both correct and incorrect. The true/false questions offered a number of statements that were either correct or incorrect in terms of properties of parallel lines. (For example, "parallel lines can cross" is false, whereas "parallel lines have to stay the same distance apart" is true.) To respond to these questions, students had to match a verbal statement with their existing conceptions about parallel lines. We anticipated that the experimental group would do better than the control group on these questions, but this was not the case. Clearly, some aspect of these questions proved equally difficult for both groups and this could be linked to the verbal nature of the test.

On the one hand, it was apparent that our experimental group generally had performed better than the control group in the long term, although for some individual students our teaching strategies had not resulted in any sustained change in their thinking about parallel lines. There is some evidence to suggest that some students did perform well on the parallel lines tests during the short-term posttest but regressed over the long term (Mansfield & Happs, 1991). It is possible that some students did not construct new meaning from our teaching strategies but continued to use their original conceptual frameworks about parallel lines.

CONCLUSION

This series of investigations has raised more questions about the complex nature of learning than it has answered. We started our investigations because of our concern that traditional teaching strategies do not bring about long-term conceptual change, and because this generally is not appreciated (as indicated by the way in which evaluation outcomes usually are expressed in terms of short-term learning gains). Our teaching strategies were based on recognition of students' alternative conceptions and the role that students' existing views have in making sense of and constructing understanding from new information, group interaction, and the articulation of ideas. Evaluation of our teaching strategies demonstrated that they generally were successful in bringing about short-term gains in understanding and retention, but that the results were not as clear-cut when long-term gains were considered.

It is interesting and useful to speculate about the reasons for our partial success with strategies that appeared to have sound theoretical underpinning. We might need to consider more closely a number of factors before we develop and evaluate further materials and teaching strategies in other areas of geometry. For example, we have little information concerning the origins of student conceptions and beliefs that might impinge on learning outcomes. Nor do we have any information about specific interactions between student beliefs and new information that is presented in the classroom. We have considered the point in mathematics education at which students construct understanding, assimilate new ideas, and are able to articulate those new ideas. It might be that this point is not reached until students have to express their new understanding to others.

One factor that might determine whether teachers will have a real commitment to using our materials and teaching strategies as intended could be their personal belief systems about how mathematics should be taught and the importance of various topics in mathematics (Cooney, 1985; Frank, 1990; Wood & Cobb, 1991). Teachers will make judgments about the effectiveness of our materials and strategies relative to their own teaching approach for this topic. Mathematics teachers themselves might not develop a total commitment to new topics in mathematics until they actually have to teach that topic. It is during this metacognitive preparation for teaching that existing conceptual frameworks related to a topic are closely scrutinized and weighed against existing conceptual frameworks that are competing with new information. Teachers may need to reconcile these competing frameworks before articulating their accepted framework to their students.

If teachers come to understand new material by the reprocessing and articulation of information, then it is possible that students also might be able to construct greater understanding of new material if they are encouraged to use the same strategies. We are working to develop and try out peer teaching strategies that will offer an environment in which students can consider new information and reconcile such information with their existing frameworks. Then, they can convey such new information to their peers in the classroom. We might have some reason for guarded optimism over our partial success with our constructivist approach to long-term conceptual change. However, we acknowledge that the complexity of the teaching and learning process could mean that no one strategy will lead to long-term conceptual change for all. We are considering the limiting factors that prevent long-term conceptual change for some students and we are in the process of modifying our teaching strategies accordingly. We are aware that, while a specific teaching strategy might challenge and change one student's thinking in mathematics, that same teaching strategy (for reasons that are still unclear) might make little impression on the entrenched views held by another student.

REFERENCES

Carpenter, T. P., & Fennema, E. (1988). Research and cognitively guided instruction. In E. Fennema, T. P. Carpenter, & S. J. Lamon (Eds.), *Integrating research on teaching and learning mathematics* (pp. 2–19). Madison: Wisconsin Center for Education Research, University of Wisconsin.

Cobb, P. (1988). The tension between theories of learning and instruction in mathematics education. *Educational Psychologist, 23,* 87–103.

Cobb, P. (1990). A constructivist perspective on information-processing theories of mathematical activity. *International Journal of Educational Research, 14,* 67–92.

Cooney, T. J. (1985). A beginning teacher's view of problem solving. *Journal for Research in Mathematics Education, 16,* 324–336.

Frank, M. L. (1990). What myths about mathematics are held and conveyed by teachers? *The Arithmetic Teacher, 37*(5), 10–12.

Fuys, D., Geddes, D., & Tischler, R. (1984). *An investigation of the Van Hiele model of thinking in geometry among adolescents: Final report.* New York: Brooklyn College.

Happs, J. C., & Mansfield, H. M. (1989, March). *Students' and teachers' perceptions of the cognitive and affective outcomes of some lessons in geometry.* Paper presented at the annual meeting of the American Educational Research Association, San Francisco, CA.

Kempa, R. F. (1993, January). *Science education research and the practice of science education.* Keynote address presented at the International Conference on Science Education in Developing Countries: From Theory to Practice, Jerusalem, Israel.

Mansfield, H. M., & Happs, J. C. (1987). Students' understanding of parallel lines: Some implications for teaching. In J. D. Novak (Ed.), *Proceedings of the Second International Seminar on Misconceptions and Educational Strategies.* Ithaca, NY: Cornell University.

Mansfield, H. M., & Happs, J. C. (1989, July). *Using concept maps to explore students' understanding in geometry.* Paper presented at the thirteenth annual conference of the International Group for the Psychology of Mathematics Education, Paris, France.

Mansfield, H. M., & Happs, J. C. (1991, July). *Difficulties in achieving long term con-*

ceptual change in geometry. Paper presented at the fifteenth annual conference of the International Group for the Psychology of Mathematics Education, Assisi, Italy.

Mansfield, H. M., & Happs, J. C. (1992). Using grade eight students' existing knowledge to teach about parallel lines. *School Science and Mathematics, 92,* 450–454.

Osborne, R. J., & Wittrock, M. C. (1983). Learning science: A generative process. *Science Education, 67,* 489–508.

Peterson, P. L. (1988). Teachers' and students' cognitional knowledge for classroom teaching and learning. *Educational Researcher, 17*(5), 5–14.

Resnick, L. B. (1989). Introduction. In L. B. Resnick (Ed.), *Knowing, learning and instruction* (pp. 1–24). Hillsdale, NJ: Erlbaum.

Vinner, S., & Dreyfus, T. (1989). Images and definitions for the concept of function. *Journal for Research in Mathematics Education, 20,* 356–366.

von Glasersfeld, E. (1987). Learning as a constructivist activity. In C. Janvier (Ed.), *Problems of representation in the teaching and learning of mathematics* (pp. 3–17). Hillsdale, NJ: Erlbaum.

Wood, T., & Cobb, P. (1991). Change in teaching mathematics: A case study. *American Educational Research Journal, 28,* 587–616.

CHAPTER 11

Teaching for Conceptual Change

Peter W. Hewson
University of Wisconsin–Madison, United States

The critical role that students' current knowledge plays in any intellectual activity is widely accepted (e.g., Kuhn, 1993). So, too, are the findings that there is considerable diversity in students' current knowledge, that the diversity frequently contradicts accepted views, and that much of this diversity seems to be unresponsive to instruction. These findings require explanation. A view of learning as simply the accumulation of information in a relatively passive manner seems inadequate. A more convincing account of these findings is provided by constructivism. Within the broad scope of constructivism, thinking of learning as a process of conceptual change provides some clarity to what it means when someone makes sense of new information by using his or her existing knowledge. This raises two issues: what conceptual change means, and what a view of learning as conceptual change says about teaching.

This chapter first outlines a model of learning as conceptual change, with particular emphasis given to clarifying the meaning of several key terms. Then, the relation of this model to teaching is considered by identifying two general guidelines for teaching. Next, discussion focuses on the necessity of a metacognitive component to teaching for conceptual change. Finally, consideration is given to several important factors that seem to be necessary in meeting these guidelines in normal classrooms; these factors relate to the teacher, the learner, and the classroom climate.

LEARNING AS CONCEPTUAL CHANGE

A model of conceptual change (or CCM) was developed by Posner, Strike, Hewson, and Gertzog (1982) at Cornell University in 1978–1979 and was expanded by Hewson (1981, 1982). It describes learning as a process in which a person changes his or her conceptions by capturing new conceptions, restructuring existing conceptions, or exchanging existing conceptions for new conceptions (i.e., a process of conceptual change). A key factor in the learning process is the status (Hewson, 1981) that new and existing conceptions have for the learner. Status is a measure of a learner's acceptance of, or preference for, an idea, new or old. Its technical details are discussed below. The model predicts that conceptual changes do not occur without concomitant changes in the relative status of changing conceptions.

When thinking of conceptual change, it is helpful to recognize that the same word, "change," can be used in different ways. One might talk, as in the fairy tale, of a frog changing into a prince when the princess kisses the frog. In this case, there is only one entity before and a different one after; the frog is no more after the change, and there is only a prince. Here change means extinction of the former state. A second example might be Jane's savings account. Her money earns interest and the balance grows; she spends money and the balance falls. Here change means an increase or decrease in the amount of something. A third example might be an election for political office with the incumbent being beaten by the challenger: There has been a change of mayor. Both people continue to live in the city, but only one person is mayor. The incumbent loses status, while the challenger gains it. In this case, there is no extension; change means an exchange of one entity for another.

Interest in conceptual change, to a considerable degree, has been focused on the problem of students who hold one view (e.g., that a table supports a book by being in the way), in contrast to a teacher's view (e.g., a table supports a book by exerting a force on it). How should one characterize a student who changes his or her mind from the former to the latter? Change as in the first example above—extinction—does not seem to be an appropriate characterization of this change. There is no sense in which one view has metamorphosed into the other; by and large, students will remember both views and simply say that "I changed my mind" or "It made more sense." Change as in the third example—exchange—seems a much better characterization of what the student has reported. It is change of this kind that is evoked for most people on hearing the term "conceptual change." It also has been called "accommodation" and "conceptual exchange."

But what of change as in the second example—increase or decrease? This seems to be a helpful characterization of learning without difficulty. Much of what students do is to learn things they didn't know by making connections to what they already know; this is not a problem when students' present views are consistent with what they learn. It has been called "assimilation" and "conceptual capture." What, then, is conceptual change? Does it refer only to instances of conceptual exchange? Or should instances of conceptual capture also be included? In this chapter, the more inclusive definition is used.

There are two major components to the CCM: the *conditions* that need to be met (or no longer met) in order for a person to experience conceptual change; and the person's *conceptual ecology*, which provides the context in which the conceptual change occurs, influences the change, and gives it meaning. The conceptual ecology consists of many different kinds of knowledge, with some of the important kinds being epistemological commitments (e.g., to consistency or generalizability), metaphysical beliefs about the world (e.g., the nature of time), and analogies and metaphors that might serve to structure new information.

The conditions apply to conceptions that a learner either holds or is considering. A critical point is that it is *the learner* who has to decide, implicitly or explicitly, whether they are met. First, is the conception *intelligible* to the learner? That is, does the learner know what it means? Is the learner able to find a way of representing the conception? Second, is the conception *plausible* to the learner? That is, if a conception is intelligible to the learner, does he or she also believe that it is true? Is it consistent with and reconcilable with other conceptions accepted by the learner? Third, is the conception *fruitful* for the learner? That is, if a conception is intelligible to the learner, does it achieve something of value for him or her? Does it solve otherwise insoluble problems? Does it suggest new possibilities, directions, ideas? The extent to which the conception meets these three conditions is termed the *status* of a person's conception. The more conditions that a conception meets, the higher is its status.

A central prediction of the CCM is that conceptual changes do not occur without concomitant changes in status. Learning a new conception means that its status rises (i.e., the learner understands it, accepts it, and sees that it is useful). If a learner sees that a new conception conflicts with an existing conception (i.e., one that already has high status for the learner), he or she cannot accept it until the status of the existing conception is lowered. This happens only if there is reason to be dissatisfied with it. The learner's conceptual ecology plays a critical role in determining the status of a conception because, among other things, it provides the criteria in terms of which he or she decides whether a given condition is (or is not) met. In this regard, the person's epistemological commitments (e.g., to generalizability) (Hewson & Hewson, 1984) are particularly important.

GUIDELINES FOR TEACHING FOR CONCEPTUAL CHANGE

The conceptual change model is a learning model. Learning models do not prescribe teaching but can be used to critique teaching, because suggested teaching sequences can be examined to see whether they facilitate or hinder learning outlined by a particular model. An examination process such as this can lead to the identification of general guidelines that are consistent with the model and function to eliminate what is inconsistent with it rather than prescribe what is necessary. Teaching that explicitly aims to help students experience conceptual-change learning and that meets guidelines consistent with the conceptual-change model is termed "teaching for conceptual change" in this chapter. The listing of guidelines for teaching in a particular order does not imply a temporal or sequencing order. Analytically, these guidelines represent different purposes that might be achieved concurrently, depending on the chosen classroom activity.

Eliciting Different Views

In teaching for conceptual change, it is necessary that the range of views related to the topic held by different people in the class be made explicit. These views need to be contributed by both teacher and students. In the process, people will become aware of ideas they previously had not encountered.

One part of this guideline is found in all normal teaching, because teachers always have made their views explicit in teaching the goals of the topic. However, there are two significant differences from common practice. The first is the consideration of students' views. This step has been advocated widely in recent literature, but still it is not the case in many classrooms. When only the teacher's views are stated explicitly, some students can be aware that their own individual ideas are different from those of the teacher, but be unaware of the views of other students. This guideline points to practice significantly different from this.

The second difference from common practice is that students' views must be considered in ways similar to the teacher's view. This might seem surprising because the teacher's view probably will be developed more completely, will be more acceptable to the science community, and is the intended learning outcome of teaching. The intent of this aspect of the guideline is to ensure that students choose between different views on the basis, not of who said them, but of how good an explanation each provides. There are implications of this point for teachers, students, and classroom climate that are considered below.

One way in which to make different views explicit involves having a teacher start a topic with a noncredit quiz, answered individually, that includes questions having a range of options that represent common different views. After the quiz, the teacher identifies the range of answers in the class and asks students to explain their choices (Minstrell, 1982). Another approach involves having a teacher provide some demonstrations to introduce the context of the work. Students individually work through elicitation activities, discuss their responses in pairs, prepare posters in groups of four to summarize their findings, and present their posters to the class (Children's Learning in Science [CLIS], 1987).

Changing the Status of Some Views

With different views to consider in teaching for conceptual change, students have to make choices. Possible outcomes might be a continuing preference for students' prior views, an acceptance of more than one view, or a preference for a different view at the expense of prior views. In this process, students are likely to find that some views become more acceptable and others less acceptable. In other words, the status of these views changes, with the status of some being raised and the status of others being lowered. In certain cases, these changes can be interdependent (e.g., when students change their minds about two views that are mutually contradictory).

It is necessary to recognize that, when a choice is made, it depends not only on the available options for choice, but also on the criteria for making the choice. These criteria, and the knowledge required to apply them, are part of each person's conceptual ecology. Because there is likely to be considerable variation between conceptual ecologies, it is likely that different people will make different choices. This notion of choice is tied into the guideline—a person chooses one option over another because of its higher status to him or her. In other words, people use their conceptual ecology in making status determinations; these probably are implicit and are stated in status language only on special occasions, but nevertheless they are critical. For instance, students reject the view that a table exerts an upward force on a book to support it because they cannot imagine how "a dead table knows how much to push up" (Hennessey, 1991). This view has low status for students who know what it means but don't believe it. In other words, a need for explanations to provide acceptable causal mechanisms is an important component of students' conceptual ecology. Other students accept this view because they explain the book's state of rest by using balanced forces, an explanation seen in other examples. In other words, a student's epistemological commitment—similar examples require similar explanations—is instrumental in raising the status of this view and is the criterion used in making a choice. The guideline therefore suggests that a teacher must be aware of the importance of the status of students' views and of components of the conceptual ecology, and include both status and ecology considerations explicitly in classroom teaching (Hennessey, 1991; Hewson & Thorley, 1989).

Activities aimed at raising the status of particular ideas always will be a part of teaching for conceptual change. In this respect, it has much in common with normal teaching. These activities might involve presenting and developing the ideas, providing examples of them, applying them in other circumstances, giving different ways of thinking about them, linking them to other ideas, and so forth. Activities aimed at lowering the status of other ideas also will be a part of teaching for conceptual change. These might involve exploring their unacceptable implications, considering experiences that they are unable to explain, or finding ways of thinking about them that point to their inadequacies. Whether any status-lowering activity works for a particular student requires that the student see the inadequacy of an idea; a common problem is that teachers often mistakenly assume that the discrepancy is as obvious to others as it is to them. Status-raising and status-lowering activities, of course, can always overlap.

In many classrooms, teachers' practices do not lead to different views' being considered. Thus, actions to facilitate students' lowering the status of their prior views do not become an explicit part of classroom practice. Without the possibility of this being so, teaching for conceptual change cannot occur. It is important to note that it is not necessary that teaching for conceptual change occur in order for conceptual-change learning to happen; rather, the role of teaching for conceptual change is that of a catalyst.

METACOGNITIVE NATURE OF TEACHING FOR CONCEPTUAL CHANGE

Teaching for conceptual change is explicitly metacognitive. The concept of metacognition refers to "knowledge concerning one's own cognitive processes and products" (Flavell, 1976, p. 232). Metacognition's cognitive process aspects have been accentuated in studies of other areas (e.g., reading), where it refers to the knowledge and control of factors that affect learning activity such as knowledge of oneself as a learner, the demands of the learning task, and the strategies employed in learning (Palincsar & Ransom, 1988). While these aspects are important for any forms of learning, knowledge of one's own cognitive products is particularly important in teaching for conceptual change. Thorley (1990) employed a helpful distinction, using the term "metaconceptual" to refer to "reflection on the content of conceptions themselves" (p. 116). In other words, students are being metaconceptual when they not only think *with* their ideas relating to a phenomenon, but also think *about* these ideas.

Why are metacognition, in general, and metaconception, in particular, hallmarks of teaching for conceptual change? When teachers elicit different explanations of a particular phenomenon or a set of phenomena in a classroom, in effect they are laying out the explanations themselves as objects of cognition; this is being metaconceptual. When students comment on, compare, and contrast these explanations, when they consider arguments to support or contradict one or another explanation, and when they choose one of the possible explanations, they are engaged in metaconceptual activities. Metaconception thus lies at the heart of the guidelines discussed above.

FACTORS SUPPORTING TEACHING FOR CONCEPTUAL CHANGE

In teaching for conceptual change, it was asserted that different views of students must be elicited, that the status of some students' views might have to change, and that such teaching is explicitly metacognitive. To accomplish this kind of teaching, the different roles that teachers and students might play and the kind of classroom climate that has to be established are described below.

The Teacher

In teaching for conceptual change, teachers have different roles to play. One role is to be the classroom manager who is responsible for establishing the classroom climate outlined below in order to facilitate student learning. This entails setting appropriate contexts for classroom activities, posing problems that have relevance and meaning to the students, exploring what underlies different ideas without threat to those who hold them, finding ways of helping students become dissatisfied with their own ideas, and introducing tasks in which students apply newly acquired ideas. It also requires that teachers set the ground rules governing all aspects of classroom interaction, discuss them explicitly with the class, and apply them consistently.

A second role is to be an active participant in the classroom, and this presents a major dilemma. On one hand, it is very easy for the teacher's voice to be the most powerful one in the class; in many classes, it is the only one. On the other hand, the teacher can employ discovery learning, with the assumption that all information comes from experience (i.e., the teacher's content voice is not heard). It is necessary to strike a balance between these two positions; it is as important to hear the teacher's view as it is to hear the views of the students. It also is as important that students feel as free to reject their teacher's views as to reject their classmates' views.

Playing these roles successfully requires important teacher characteristics, some of which have been discussed in the previous sections: a respect for and knowledge of learners and their ideas, and a wide repertoire of appropriate teaching strategies and supporting materials. Implicit in all of these are the teacher's conceptions of the nature of learning, of teaching, and of science that are supportive of teaching for conceptual change. These have been outlined elsewhere (Hewson & Hewson, 1988).

The Learner

In teaching for conceptual change, students have particular roles. They must become learners who are convinced that the goal of learning should be to understand the topic being considered and, in doing so, to make it their own. Thus, students have to accept responsibility for their own learning, trust their own thinking, and justify their conclusions using sensible arguments. In doing so, however, they must recognize that there might be different views of the same event and that those views must be respected. When different views are expressed, students must listen to and understand these views, and negotiate common meanings. Finally, students should be prepared, in light of the comparison and contrast of views, to change their view when another seems to be more viable.

Classroom Climate

The climate in a classroom where teaching for conceptual change occurs has several significant features. As mentioned above, it is the teacher who must establish the classroom climate. Both teacher and students must respect the ideas of others in the classroom and listen carefully to them, even though they might not agree with them. It is essential for participants, without fear of sanction or ridicule, to be able to express their ideas openly, to express their disagreement with the ideas of others, and to ask for clarification of the explanations of others. Another dimension is the need to separate person and idea: to be able to critique an idea while affirming the person. One strategy for achieving this is to hide the identity of the source of an idea (e.g., by eliciting ideas in groups, collect-ing written ideas anonymously, or role-playing other peoples' ideas).

Next, there has to be a common acceptance that the goal of discourse is the achievement of shared meanings about the topics under discussion. Because the diversity of initial views in the classroom must be recog-nized, achieving this goal requires a willingness to make the effort to understand others' points of view, to negotiate, and to compromise. This is a time-consuming process that can easily be subverted by tactics such as the premature closure of debate. Finally, the negotiated meanings must be adopted because they make sense to the participants and not because the teacher said so. In other words, the basis for acceptance should be the rationality of the topic under consideration rather than the authority of the source of the accepted meanings, be it textbook, teacher, or individual student. An important part of achieving this is the even-handed consider-ation of views outlined in the guidelines above.

Although these features must be operative at all times in the class-room (i.e., during the elicitation of different views and during status-changing activities), they are not common practice in many classrooms. Frequently, there is only one view for consideration, that of the teacher. The implicit assumption is that students' views are copies (probably im-perfect) of the teacher's and are otherwise of no consequence. The teach-er's view is transmitted to the students, and the only negotiation centers on whether the students have received this view, regardless of whether it makes sense to them.

CONCLUSION

This chapter has outlined a model of learning as conceptual change and has described two general guidelines for such teaching to occur. In conceptual-change teaching, which is highly metacognitive, there is need for eliciting different student views and probably for the status of some student views to change. This kind of teaching is accomplished when teachers, students, and the classroom climate take on features that sup-port teaching for conceptual change. While these features might be neces-sary, certainly they are not sufficient. They are not complete because nei-ther curriculum nor assessment has been mentioned. Also, they are not detailed enough; putting them into practice requires the development of many different activities related to the topic being considered and the ideas that students bring with them.

Each of these features in isolation might not seem very different from current teaching practice. After all, dedicated teachers over the years have developed science activities and employed teaching strategies that have challenged many students and facilitated much science learning; without doubt, teaching for conceptual change benefits from this huge base of expertise. Yet, when all of these features are combined successfully, the change is far from incremental. In comparison with most current teaching, teaching for conceptual change is a radically different enterprise.

REFERENCES

Children's Learning in Science (CLIS). (1987). *Approaches to teaching the particulate theory of matter.* Leeds, England: University of Leeds.

Flavell, J. H. (1976). Metacognitive aspects of problem solving. In L. B. Resnick (Ed.), *The nature of intelligence* (pp. 231–235). Hillsdale, NJ: Erlbaum.

Hennessey, M. G. (1991). *Analysis of conceptual change and status change in sixth graders' concepts of force and motion.* Unpublished doctoral dissertation, University of Wisconsin–Madison.

Hewson, P. W. (1981). A conceptual change approach to learning science. *European Journal of Science Education, 3,* 383–396.

Hewson, P. W. (1982). A case study of conceptual change in special relativity: The influence of prior knowledge in learning. *European Journal of Science Education, 4,* 61–78.

Hewson, P. W., & Hewson, M. G. A'B. (1984). The role of conceptual conflict in conceptual change and the design of instruction. *Instructional Science, 13,* 1–13.

Hewson, P. W., & Hewson, M. G. A'B. (1988). An appropriate conception of teaching science: A view from studies of science learning. *Science Education, 72,* 597–614.

Hewson, P. W., & Thorley, N. R. (1989). The conditions of conceptual change in the classroom. *International Journal of Science Education, 11,* 541–553.

Kuhn, D. (1993). Science as argument: Implications for teaching and learning. *Science Education, 17,* 319–337.

Minstrell, J. (1982). Explaining the "at rest" condition of an object. *The Physics Teacher, 20,* 10–14.

Palincsar, A. S., & Ransom, K. (1988). From the mystery spot to the thoughtful spot: The instruction of metacognitive strategies. *The Reading Teacher, 41,* 784–789.

Posner, G. J., Strike, K. A., Hewson, P. W., & Gertzog, W. A. (1982). Accommodation of a scientific conception: Toward a theory of conceptual change. *Science Education, 66,* 211–227.

Thorley, N. R. (1990). *The role of the conceptual change model in the interpretation of classroom interactions.* Unpublished doctoral dissertation, University of Wisconsin–Madison.

CHAPTER 12

Contrastive Teaching: A Strategy to Promote Qualitative Conceptual Understanding of Science

Horst Schecker and Hans Niedderer
University of Bremen, Germany

Since the 1980s, the Institute of Physics Education at the University of Bremen has carried out empirical research on students' alternative conceptions in mechanics (Schecker, 1985), atomic physics (Bethge, 1988), and science philosophy (Meyling, 1990), and this has been based on the theoretical background explicated in Niedderer and Schecker (1992). Development and trials of new teaching strategies have been related closely to this work, as our aim has been to draw conclusions from our findings about students' ideas for the creation of better learning environments.

Following basic constructivist principles, the teaching strategy described in this chapter enables students to elicit their ideas and express them freely in class before a new science concept is introduced by the teacher. We use the term "contrastive teaching" in a learner-directed approach because of parallels with contrastive grammar, a linguistic method for teaching/learning a foreign language in which grammatical features of the target language are introduced by comparing them explicitly with related structures of the mother tongue. When considering teaching, the students' intuitive ideas about scientific phenomena correspond to the mother tongue; scientific views and concepts correspond to the target language.

We hypothesize that, as long as students are not aware of their intuitive notions, they will hardly be able to learn a related scientific concept. For example, students believe that "force" is an easy-to-learn concept because its meaning seems to be obvious from everyday experiences (Schecker, 1985, p. 452). However, empirical studies suggest that force is one of the most difficult concepts to learn—primarily because students think it is so simple. Considerable teaching effort is required to help students notice the differences between their intuitive views derived from everyday experience and the scientific view based on theory-laden observations. The constructivist view accentuates the student's active role in the learning process and implies that the teacher will create learning conditions for active engagement in learning. The learning environment that is important for contrastive teaching includes a focus on qualitative understanding, a means of handling students' alternative results, attention to students' ideas about physics teaching, and an appreciation of students' ideas.

In this chapter, we describe a case study of a student-oriented teaching episode in mechanics that exemplifies how contrastive teaching can contribute to student understanding of "force" and how this understanding is differentiated from "energy." Initially, we describe the six phases of the contrastive teaching strategy.

A CONTRASTIVE TEACHING STRATEGY

Our contrastive teaching strategy, first published in Niedderer and Schecker (1982), presupposes a certain capability of meta-analysis on the learner's side and includes epistemological questions. It is meant mainly for the upper secondary level (students age 16–19 years). Driver and Oldham (1986) propose a similar strategy for younger students, in which the "elicitation of ideas" and the "input of scientific view" are similar to our stages 3 and 5. The contrastive teaching strategy can be broken up into six stages of (1) preparation, (2) initiation, (3) performance, (4) discussion of findings, (5) comparison with scientific theory, and (6) reflection.

The first stage (*preparation*) typically involves conventional teaching with demonstration experiments and teacher-dominated presentation of concepts that precedes the start of the contrastive teaching strategy. In the second stage (*initiation*), an open-ended problem is posed by the teacher, who sketches a broad framework for students' activities (e.g., "What does acceleration depend on?"), offers a set of apparatuses for open-ended experiments, or shows an initial experiment without explaining it. The students work out questions and hypotheses for their own investigations. In stage 3 (*performance*), students perform experiments, calculations, and derivations and formulate the results in their own words. The teacher does not interfere with the students' activities but acts as a counselor, helps reservedly with technical problems, and supervises an organized working process by encouraging students to write down questions, ideas, intermediate results, and findings.

During stage 4 (*discussion of findings*), the student groups present their results in a class forum where the teacher writes notes on the board, using the students' words. The students compare their findings and try to arrive at common conclusions. The teacher challenges the students' ideas by indicating inconsistencies or suggesting additional experiments. The students defend their notions, and perhaps modify them slightly. However, this stage usually does not change students' ideas immediately. During stage 5 (*comparison with scientific theory*), the teacher brings in the scientific explanation (concepts, principles, law) as an *alternative view* to the students' ideas—not as "the truth"—and compares this with students' ideas from the preceding stage. Commonalties and differences are made explicit. The teacher shows advantages of scientific theory for universal application and precise predictions in a controllable setting. Intuitive conceptions are described as more appropriate and better suited for everyday communication about specific single events. During stage 6 (*reflection*), students look back on their problem-finding and problem-solving processes and consider methodological and epistemological issues. Findings from the philosophy of science about the different structures of everyday-life thinking and scientific thinking can help students to notice and accept these differences.

Examples for the application of this strategy, which can last from a few minutes to several weeks, are given in Niedderer and Schecker (1982), Schecker (1985), and Niedderer (1987). A teaching episode need not include all six stages. A short unit can consist of just an open-ended demonstration experiment for which the students write down their observations and questions before any explanations are given by the teacher. Contrastive teaching is not meant to be the overall strategy used in a class. Longer units like the one described below should take place once or twice a semester.

CASE STUDY: UNDERSTANDING FORCE

This case study of contrastive teaching and learning was carried out in an 11th-grade advanced physics course over two weeks (10 lessons for stages 1 to 6). The aim was to motivate students to test and further develop their ideas on force by means of self-developed problems from the domain of "collisions."

Preparation

Newton's laws of motion were introduced: for example, experiments on an air track were carried out by the teacher to derive $F = m \cdot a$. Several textbook problems with calculations of velocities, distances, and time intervals were posed.

Initiation

During stage 2, the teacher proposed the *investigation of collisions* as a frame topic. The students were prompted to form groups and, in about 45 minutes, formulate precise questions in this field together with appropriate experimental settings that could help them to find the answers. All materials in the physics laboratory were at their disposal and students were asked to list all ideas that came into their minds. Students were assured that the evaluation of their work would depend not on formal conformity with textbook explanations, but rather on the creativity and internal plausibility of their own results—even if they deviated from the textbook. Their reports were to show the steps taken in the investigations and not just the results. At the end of the initiating lesson, the groups presented their ideas in a class forum. A typical idea from students is given below:

> We want to investigate various effects of impact with cars (e.g., that one car hits the other into its side or head on—or that one of the cars also has a velocity when the other one hits it). . . . We would like to measure the *force of impact*. And we thought that we could quantify this somehow—we believe that the force can be calculated). . . . We want to measure the *transfer of force*—how much energy is left after the collision.

Two types of investigations were common in the students' spontaneous ideas as well as in the subjects finally chosen: investigations of the transfer and preservation of force/energy, and investigations of the impact force of/on a moving body.

Performance

During stage 3, the activities of all the groups centered around the questions mentioned above. The underlying ideas of force were that moving bodies have force that is actualized during the impact and is transferable to other bodies. This force can be measured from the resulting effects (i.e., either from the velocity given to the body pushed or from its plastic deformation). Time intervals, so essential for Newton's definition of force, were not considered by any of the groups. Newton's definition of force had almost no significance for the choice of subjects and the problem-

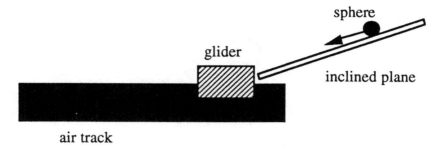

air track

Figure 12.1. Experimental Arrangement Used by Arnim and Ulf to Examine Whether the Formula $F = m \cdot a$ Is Suitable to Quantify "Force"

solving processes (except for group 5 as shown below). From the preceding instruction about inertia and the interrelation of force and acceleration, only the equation $F = m \cdot a$ was taken up. Some of the groups tried to quantify the force of moving bodies by connecting this "formula" to their intuitive understanding of force related to energy.

All six groups produced extended reports about their ideas and experiments. The students were requested to document the origin of their questions and intermediate ideas as well as the results. They developed many creative ideas to find answers for their self-defined problems. Group 5 worked on a problem directly referring to the preceding unit on Newton's laws (see Figure 12.1 for group 5's experimental setup). The students examined whether the formula $F = m \cdot a$ was really suitable to quantify all aspects of what they understood to be force, and made the following comments:

> We wanted to calculate the force that has to be exerted on a glider on the air track so that it gains a certain measurable velocity. We then thought that the force of the sphere rolling down the inclined plane is described by $F = m \cdot a$. By calculating the force and measuring the velocity, we could interrelate these two aspects. But this of course was complete nonsense as the F in the formula is the force acting on the sphere and not the one exerted by the sphere. We thus asked the question whether the formula still could be valid for the force exerted by the sphere. However, we found contradictions in two different ways.
>
> First, deliberation gave rise to contradiction. The mass of the sphere is always the same, so the force exerted by the sphere also would have to be always the same. It therefore would be of no sig-

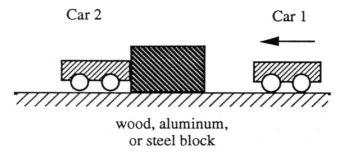

Car 2   Car 1

wood, aluminum,
or steel block

Figure 12.2. Experimental Arrangement Used by Markus and Andreas to Examine How a "Force" Acts Through a Heavy Obstacle

nificance from which height we let the sphere roll down. The glider always would gain the same velocity as the acting force was always equal. This seemed absurd from pure logic.

Second, contradictions arose from experiments. The sphere rolling from different heights resulted in different velocities of the glider. We then wanted to relate the forces and therefore needed the correlation of the accelerations. But, for the glider achieving an acceleration from 0 to 100 in practically no time (first it stands, then it runs at a constant velocity), we could not state any acceleration on the air track.

The train of thought is described logically. Arnim and Ulf found out in the end that *acting forces* are deduced from accelerations. The force exerted from the sphere on the glider could be calculated from the acceleration of the glider, which, however, is not measurable. The students further found out that the "force exerted (or exertable) by the sphere" and the force "acting on the sphere" are different things and that $F = m \cdot a$ is valid only for the latter case.

Markus and Andreas (group 6) chose the problem "How does a force act through a heavy obstacle?" A car coming from the right hits against a block behind which a second car stands. The velocity (or speed) of car 1 before the push, and that of car 2 after the push, were thought to give information about the transfer of force or loss of force, respectively. Figure 12.2 shows group 6's experimental setup. One of the students in group 6 made the following comments:

To investigate where the force was going, we wanted to calculate the force of car 2 after the push by the formula $F = m \cdot a$, with the acceleration being derived from $[(\Delta s / \Delta t) / \Delta t]$. Even before we knew from where to take t, we saw our first error. We could not calculate the forces of the cars by $F = m \cdot a$ as the cars did not accelerate at all but rather had a uniform velocity.

We postponed the question of whether force could be calculated in some other way or whether it was really force that the cars had, and tried to draw some conclusions from the velocities. . . .

Conclusions: Acceleration results from a force acting continuously. In this sense, there is force. In the case of uniform motion, the force acts only once—as a starter. Afterwards force is no longer present. Instead, a body moving at constant speed must have energy that becomes noticeable when it hits another object. This energy must be proportional to mass and velocity. We call it kinetic energy.

In concluding their ideas, Markus and Andreas differentiated the idea of *force* to obtain acceleration of a body and *energy* of the moving body. As in Arnim and Ulf's work, the cluster concept of force started to break up when the students had opportunities to develop their own ideas over a longer time and to test their definition of force on a subject of their own choice. The awareness of differences was triggered particularly by the formula $F = m \cdot a$, which proved to be inappropriate for quantifying what they understood to be force.

Discussion of Findings

Stage 4 occurred within the general subject of collisions, when all groups had worked on problems around the definition of force. After the groups reported their investigations and conclusions in a class forum, the teacher listed some central statements on the board and tried to keep to students' words. An excerpt from the transcript is given below:

TEACHER: On the board, we very often find the word "force": force of impact, starting force, force of the car, exerted force, force is transferred. Does the word force always refer to the same thing?

MARKUS: In our case, I would not define what we had with uniform motion as force because, during acceleration, the force is acting continuously. Therefore it is presumably present. In case of a uniform velocity, the force acts only once as a starter for velocity and then no longer can be in the car. I therefore would describe the rolling car in our case—what is in it—to be kinetic energy.

ARNIM: For example, I would say—concerning the force exerted and the force of the car—that the car has some sort of force and can transfer this afterwards. And, in other cases, it's the force that's exerted. That's the difference.

CLAUS: I would say that the car running along the track has kinetic energy. I always call that energy. Then the kinetic energy is transferred into deformation energy—if you can say so—relating to the spring [which is fitted to the experiment vehicle]. This spring transfers this energy again to another spring, resulting again in kinetic energy and with energy being transformed again and again.

The students proposed a distinction between several types of forces (Arnim), or used the different terms "force" and "energy" (Markus, Claus). The discussion had reached a state similar to the situation of physics in the second half of the nineteenth century. This was a chance to introduce an extract from a Helmholtz text, published in 1861, that calls for a distinction between the "intensity of force" (today *force*) and the "total amount of driving force" (an aspect of today's *energy*).

Comparison with Scientific Theory

During stage 5, original texts are presented only *if* and *after* the students have elaborated a similar approach. They are treated not as historical documents ("What did Helmholtz think about . . .") but as illustrations of views that are still virulent in today's thinking. Historical discussions thus can play an important role in helping students to develop conceptual awareness (Schecker, 1992). Students can be encouraged to do self-directed work when they learn that results differing from the textbook have been held by famous scientists.

The original text by Hermann von Helmholtz, entitled *About the Application of the Law of Conservation of Force ("Kraft") on the Organic Nature* (abridged from Samburski, 1978), is presented below:

> The most distinguished progress of the natural sciences in our century was the discovery of a universal law which covers and governs all the different branches of the physical and chemical sciences. This law is today called "the principle of conservation of force." A better denomination might be that of Mr. Rankine who speaks of "conservation of energy" because the law does not relate to what we normally call intensity of force. It does not mean that the intensity of natural forces is constant but rather relates to the total amount of driving force won by some natural process, by which a certain amount of work can be done. . . . There is, however, still another type of mechanical driving force, and that is velocity. If the velocity of a body does work we call it vis viva or living force of a body. The enormous force of a canon ball only depends on its velocity. (p. 518)

The teacher had prepared the text as a handout because he expected, from research into students' ideas about force, that the class probably would arrive at similar ideas. After reading the text and clarifying its content, the teacher asked questions. Some excerpts from a longer transcript of the dialogue between teacher and students are:

TEACHER: Can you draw any conclusions from this text for your experiments? Or can't you see any connections?

LORENZ: Helmholtz uses alternating concepts. He does not decide.

TEACHER: What do you mean by "he does not decide"?

LORENZ: At one point, he speaks of force that is transformed, . . . and then he speaks of force that is exerted. I don't really get that. This other guy, Rankine, suddenly speaks of energy. He speaks of energy instead of force. That's where the two differ. The concepts are not defined clearly.

TEACHER: Helmholtz proposes rather to speak of conservation of energy instead of the principle of conservation of force. Does he keep to his own proposal?

ULF: No, he doesn't. He still calls it force. For example, in his last sentence ("The enormous force of a cannon ball only depends on its velocity"), I would say that it's just the energy which the bullet has when it is launched.

A physically sound understanding of force and energy can develop only in a process of mutual delimitation. It becomes clear how the students are stimulated by attempts to differentiate between Newtonian force and energy.

In this case study, stage 6 (reflection) was not a separate stage of the unit; rather parts of the reflection stage were included in stages 4 and 5. Thorough reflection on the problem solution process becomes particularly noticable in the comprehensive reports that the students produced after the 10-lesson unit, as illustrated by the quotes from groups 5 and 6 above.

CONCLUSION

This chapter describes the six stages of a learner-directed strategy called *contrastive teaching,* which is founded on basic constructivist principles. The case study involving classroom observations and students' reports concluded that the contrastive teaching/learning unit on collisions can be used to promote qualitative conceptual understanding of force. Students' confrontation with consequences resulting from the current perspective they had worked out themselves was a substantial step toward establishing the scientific view in explicit contrast to intuitive thinking. Prior to this strategy, teacher-oriented instruction had not succeeded in helping students to develop a physical understanding of force. Rather, the students simply had added formal elements of Newton's concept as another facet to their undifferentiated everyday-life force/energy/thrust preconcept.

Case studies do not provide statistical evidence for positive effects of constructivist teaching compared with conventional strategies. However, students from the class taught with the contrastive teaching strategy had significantly higher scores on a questionnaire on conceptual understanding in mechanics than the average of about 250 students taught in a conventional manner. Current research in our institute focuses on the field of quantum mechanics. Starting from the students' picture of the atom as a small solar system, contrastive teaching is employed to help students gain new views of electrons and nuclei (Petri & Niedderer, 1994).

REFERENCES

Bethge, T. (1988). *Aspects of students' ideas about basic concepts of atomic physics.* Unpublished doctoral dissertation, University of Bremen, Germany.

Driver, R., & Oldham, V. (1986). A constructivist approach to curriculum development in science. *Studies in Science Education, 13,* 105–112.

Meyling, H. (1990). *Science philosophy in upper secondary physics courses.* Unpublished doctoral dissertation, University of Bremen, Germany.

Niedderer, H. (1987). A teaching strategy based on students' alternative frameworks: Theoretical concept and examples. In J. D. Novak (Ed.), *Proceedings of the Second International Seminar on Misconceptions and Educational Strategies in Science and Mathematics* (Vol. II, pp. 360–367). Ithaca, NY: Cornell University.

Niedderer, H., & Schecker, H. (1982). Aims and methods of science-philosophy oriented physics teaching. *Der Physikunterricht, 16,* 58–71.

Niedderer, H., & Schecker, H. (1992). Towards an explicit description of cognitive systems for research in physics learning. In R. Duit, F. Goldberg, & H. Niedderer (Eds.), *Research in physics learning: Theoretical issues and empirical studies—Proceedings of an international workshop held in Bremen, Germany* (pp. 74–98). Kiel, Germany: Institute for Science Education (IPN), University of Kiel.

Petri, J., & Niedderer, H. (1994). A case study about changing student conceptions in atomic physics. In H. Berendt (Ed.), *Zur Didaktik der Physik und Chemie* [Physics and Chemistry Education] (pp. 298–300). Alsbach, Germany: Leuchtturm Verlag.

Samburski, S. (1978). *Der Weg der Physik* [The historical pathway of physics]. München, Germany: Deutscher Taschenbuch Verlag (German Pocketbook Publishers).

Schecker, H. (1985). *Students' matrices of understanding in mechanics.* Unpublished doctoral dissertation, University of Bremen, Germany.
Schecker, H. (1992). The paradigmatic change in mechanics: Conclusions from the history of science for physics education. *Science and Education, 1,* 71–76.

CHAPTER 13

Concept Substitution: A Strategy for Promoting Conceptual Change

Diane J. Grayson
University of Natal, Pietermaritzburg, South Africa

A major thrust in physics education in the last decade or so has been the identification of student "misconceptions" (see Pfundt & Duit, 1994, for a bibliography). In recent years, other terms for describing students' ideas have become popular, including "naive conceptions," "alternative conceptions," and "preconceptions." I prefer these terms because "misconception" implies that the students' ideas are wrong, which might not always be the case. Some researchers have started to look at student conceptions with a view to identifying correct parts as well as "incorrect" parts (e.g., Clement, Brown, & Zietsman, 1989).

An important research area in science education generally is conceptual change. Research has shown that student ideas that conflict with accepted scientific ideas are often remarkably difficult to change by means of instruction (Minstrell, 1984). Researchers have studied the conditions and environment needed for conceptual change to occur in students who hold such ideas (Hewson & Thorley, 1989). Scott, Asoko, and Driver (1992) give an overview of documented strategies for promoting conceptual change. In particular, they identify two categories of strategies: those based on cognitive conflict and the resolution of conflicting perspectives and those that build on learners' existing ideas and extend them.

This chapter describes an instructional strategy, called *concept substitution*, for promoting conceptual change in situations in which students appear to have a correct intuition associated with an inappropriate physics term. The chapter begins with a description of the context. The use of concept substitution is then illustrated in the domains of electricity and heat and temperature. Finally, the advantages of this approach are summarized.

CONTEXTS

The instructional strategy described in this chapter falls into the second category identified by Scott and colleagues (1992) in that it builds on learners' existing ideas. In Duit's (1993) framework, in which the pathway from students' preconceptions to science conceptions can be classified as either continuous (evolutionary) or discontinuous (revolutionary), the strategy falls into the continuous-pathway class.

This strategy is appropriate when students express an intuitive idea that is correct in terms of explaining some observed phenomenon but is associated with an inappropriate physics term by students. The strategy involves reinforcing students' correct ideas, but substituting the correct physics term. In this way, students have something on which to pin their ideas, and the concept under study can be separated in their minds from the intuitive idea. I call this strategy concept substitution because the *instructor* substitutes the concept that matches the students' intuitive ideas for the term they used incorrectly. The process allows students to embrace the accepted scientific concept without having to suspend their intuitions; intuitions are relabeled. Concept substitution is similar to Jung's (1986) notion of "re-interpretation" in that correct aspects of students' thinking are identified and used as a starting point for the development of scientific conceptions.

This chapter describes the use of concept substitution with students in a foundation physics course. The students were enrolled in the Science Foundation Program (SFP) at the University of Natal, either in 1991 (30 students) or in 1992 (32 students). The students were a mix of male and female, young (17 years) and older (up to 33 years), recent school-leavers and those who had worked for a while, and urban and rural. The SFP is a yearlong program preceding entry into a pure or applied science degree that is designed for academically talented but underprepared black students. It was instituted in order to increase the abysmally small number of black students obtaining science qualifications in higher education. It is an integrated, holistic program comprising courses in biology, chemistry, physics, mathematics, and English language development, together with a counseling component. Students are selected on the basis of their performance during a two-week residential selection program comprising mini-courses in biology, physics, and mathematics. The selection mechanism attempts to measure students' "zone of proximal development" (Vygotsky, 1978) for the three selected science subjects.

The exercises and tests referred to in this chapter comprised part of the SFP Physics course. Although almost all of the students had studied physical science up to matriculation level, most of the learning probably had been by rote, with little or no practical work experience or conceptual understanding. Thus it would be unrealistic to assume that students had much prior functional knowledge of science. Where student answers are quoted verbatim, it should be noted that English is the students' second language (if not the third or fourth).

CONCEPTUAL DIFFICULTIES IN ELECTRICITY

Numerous studies have been conducted that identify student conceptual difficulties with electricity (Duit, Jung, & von Rhoneck, 1985). One of the most prevalent and fundamental difficulties is the idea that current is used up in a circuit (e.g., Shipstone, 1985). The sequence used in class to try to evoke documented conceptual difficulties, if present, and then to remediate them is described as follows: administration of a written pretest involving predictions about what would happen in particular circuits; class debate of pretest results with a circuit set up in front of the class (but not closed) and students voting on the outcome; demonstration of an actual circuit; and discussion of why the circuit behaved as it did. The use of concept substitution during the discussion is illustrated below.

The course was based on the "Electric Circuits" module developed by the Physics Education Group (1988a) at the University of Washington, but adapted for a lecture-practical mode of instruction. In keeping with the approach used in those materials, we postulated at the outset that the brightness of a light bulb would indicate the amount of current flowing through the bulb. In one of the pretests early in the course, students were asked to predict the relative brightnesses (and therefore the amount of current) of two bulbs in series, and also to compare the brightness of those bulbs to that of a bulb in a single-bulb circuit (see Figure 13.1).

One-third of the class (10 students) predicted that the bulb the current reached "first" would be brighter. Most students thought that bulb #1 would be brighter because current flowed from the positive to the negative terminal, but a few students thought that current flowed the other way and so they said that bulb #2 would be brighter. One student gave the following explanation:

> These bulbs are connected in series. The way in which bulb #1 will glow will be different from the way in which bulb #2 will glow. When

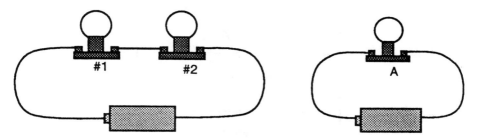

Figure 13.1. Circuits Shown in an Electricity Pretest. Students were asked to compare the brightnesses of bulbs #1 and #2 to each other and to the brightness of bulb A.

the bulbs are connected in series, they never share the current and that implies that bulb #1 will glow brightly and bulb #2 will glow dimly (i.e., most of the current is used up in bulb #1 and the little that is left passes to bulb #2).

The pretest was followed the next day by a demonstration in which students had to predict the relative brightness of three bulbs in series. There was much debate about what would happen, with one part of the class arguing that the brightness of the bulbs decreases from being brightest near the end of the battery from which the current flowed to being dimmest nearest the other end of the battery; other students said that the bulbs all would have the same brightness. When I closed the switch, the students saw that the position in the circuit did not affect the brightness of a bulb. Thus, it must follow that the current through each bulb was the same and that therefore the current could not be used up.

Concept Substitution: "Energy" for "Current"

At this point, the misconception has been evoked, confronted, and supposedly resolved. However, students who thought that current would be used up still could be uneasy. After all, batteries go flat after a while. How can the current *not* be used up? Frequently, instruction stops here, but I believe that this is a dangerous state in which to leave a student. Because they have not really resolved the issue of what was wrong with their own preconceptions, students might not make the conceptual change to believing that current is not used up, or they might appear to accept the new ideas for a time and revert to their previous idea some time later (White, 1992). This is the point at which I introduced concept

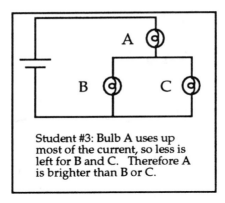

Student #3: Bulb A uses up
most of the current, so less is
left for B and C. Therefore A
is brighter than B or C.

Figure 13.2. Circuit Shown in Homework Assignment and "Explanation"
Given by a Fictitious Student

substitution. Students were told that they were *right* to say that something
is "used up," as evidenced by the fact that batteries do go flat, but that
the "something" that is used up is energy and not current. Current just
goes around and around the circuit.

In this approach, the instructor introduces the correct physics term
at the time the apparent misconception arises, even though that term
might not have been encountered before in the course. In this example,
this was the students' first exposure to the term "energy." In other words,
the term was introduced before the concept was developed. Some physi-
cists might find such apparent lack of rigor distasteful, but the advantage
of the approach is that it allows students the opportunity to transfer many
of their correct intuitions to the new term, thus disentangling intuition
from the concept that is the focus of instruction at the time (current, in
this case).

About two weeks later, I gave a homework assignment in which students were shown a circuit and asked to comment on explanations given
by four fictitious students. Figure 13.2 contains the circuit, which was
designed specifically to see whether students still held the idea that current is used up, and the "explanation" given by fictitious Student #3. Only
2 of 30 students in the class indicated that this explanation was correct
and that current was used up. The rest of the students said that current
is not used up, and some specifically referred to the fact that it is energy
rather than current that is used up, as illustrated by the following quotation: "Student #3 is right to say that bulb #A is brighter than B or C, but

wrong to say the current is used up because current is not used up but instead electrical energy is used."

CONCEPTUAL DIFFICULTIES WITH HEAT AND TEMPERATURE

The concept of temperature is a difficult one for many students, and often is confused with heat (e.g., Erickson & Tiberghien, 1985). In particular, students often believe that, if one object feels colder than another, it must be at a lower temperature than the first object. In order to test whether my students held this idea, I interviewed them concerning their ideas about heat and temperature prior to teaching the section on heat. As part of the interview, I asked them to feel a block of aluminum and a block of wood that had been sitting on the table from the time when they entered the room, and to tell me how they thought that the temperatures compared. As expected, all of the 32 students thought that the aluminum had a lower temperature than the wood because it felt colder.

About a week later, students did an experiment in which they first felt two hollow cylinders, one of wood and one of brass, with their hands and tried to compare their temperatures. They then inserted a thermometer into each cylinder and compared the readings on the thermometers (see Physics Education Group, 1988b). The students were perplexed at obtaining the same reading, and some students were sure that they had done something wrong. Again, this is the point at which instruction often stops. The students' misconception has been confronted, yet many students are left feeling confused and uneasy. Consequently, concept substitution was introduced.

Concept Substitution: "Heat Transfer" for "Temperature"

Several days after doing the experiment involving the two cylinders, I led a class discussion about the confusing results, explaining that students were correct to say that the cylinders felt different. However, that was not because they had different temperatures. Rather, what made the cylinders feel different was the rate at which each object was transferring heat away from the students' hands. I stressed that temperature is not something that can be determined using one's hand; it can be measured only with a thermometer. Thus, students could separate out the concept of temperature from what made the cylinders feel different, namely, the rate of heat transfer.

As in the electricity example, the new term "heat transfer" was introduced by the instructor before the concept was developed properly. Intro-

---

<u>Quiz</u>

Two students are in the supermarket. One of them takes a can of fish and a package of cookies off the shelf. They then have the following disagreement:

Student #1: The can of fish feels colder than the package of cookies. It has a lower temperature.

Student #2: No, you are wrong. Both the fish and the cookies have been sitting on the shelf for quite some time, so they both must have the same temperature.

Say which, if either, of the students is correct and why.

<u>Examination Question</u>

A student makes the following comments: "In my room, the floor is made of tiles; part of the floor is covered with a rug. When I walk barefoot on the tiles, my feet feel cold. But, when I walk on the rug, they feel warm. That means that the temperature of the tiles is lower than the temperature of the rug." Is this student correct? Explain why or why not.

Figure 13.3. Test Questions Involving the Concepts of Temperature and Heat Transfer

duction of the term allowed students the opportunity to associate their intuitions with this term, and so to free the concept of temperature under study from inappropriate associations. During the next class period, after a week's vacation and a week and a half after our discussion, I gave a quiz (see Figure 13.3) to the students.

Of the 32 students, only 2 gave incorrect answers to this question. Moreover, a number of the explanations showed a very good understanding of the distinction between temperature and what one feels with one's hand. The quotations below show that the students also had a good grasp of the concept of thermal equilibrium:

> Student #2 is correct because a person cannot feel an object by hand and then predict which of the objects has a higher temperature or a lower temperature. The reason why the can feels colder is that heat is being transferred from the student's hand into the can. Furthermore, the can and the package of cookies have been sitting on the shelf for a long time, and therefore they have reached thermal equilibrium with the surroundings. That means both the can and the package

of cookies have the same temperature as that of the environment in the supermarket.

What confuses Student #1 is that the can of fish feels cold, which is caused by the heat transferred from his hand to the can of fish. This is influenced perhaps by the can of fish having a higher thermal conductivity than the package of cookies. If they have been on the shelf for quite some time, they definitely will reach thermal equilibrium with the temperature of the surroundings and thermal equilibrium will mean both the can of fish and package of cookies have the same temperature.

The fact that students were able to give such good explanations on a surprise quiz after a week's vacation strongly suggests that the ideas made sense to them. I believe that, for students to give the kind of responses illustrated above without having had any chance to "cram" for the test, they must have been able to incorporate the concepts, particularly the conceptual distinction, into their own knowledge structures. Further evidence to support these claims was obtained from students' answers to an examination question (see Figure 13.3) given two months after the surprise quiz.

Each of the 32 students in the class said that the "student" quoted in the examination question was incorrect. (In the preinstruction interviews, all students thought that an object that felt colder had a lower temperature.) A typical student response is given below:

This student is not correct when he/she says that the temperature of the tiles is lower than the temperature of the rug. A person cannot compare the temperatures of two different substances. What the student is feeling is not necessarily lower temperature and higher temperature. What she is feeling is the rate at which heat is lost to the tiles or to the rug. The tiles feel colder because the rate at which heat is lost from her feet to the tiles is higher than the rate at which heat is lost from her feet to the rug.

CONCLUSION

In this chapter, several examples have been given in which an apparent misconception held by a student can be a correct intuition that has been associated by the student with the incorrect physics term. Rather than just trying to disabuse students of an idea, it is preferable to give them an alternative concept on which to hang the idea. The particular choice of sequencing of concepts described in this chapter is not original. However, the particular approach is somewhat different from other treatments. Specifically, in the process of concept substitution, a situation is created in which it is likely that students will make an incorrect association between a certain intuitive idea and a physics term. When this happens, the instructor reinforces the student's correct intuition but assigns it another label; that is, the instructor substitutes the name of the concept with which the student's idea can be associated correctly.

Some science educators might take exception to introducing a new concept "out of the blue," or at least without a careful lead-in to the concept that is being substituted. However, in the long run, this disadvantage could be outweighed by the advantage of students' being introduced, early on in the study of a new topic, to a term with which they can associate their intuitions, even if that term is not developed fully or defined when it is introduced initially. This approach allows the concept on which the instructor is trying to focus to be freed from the conceptual baggage that students might try to load onto it. Moreover, when the newly substituted concept is properly developed at a later stage, students already have some feeling for that concept.

In summary, instead of focusing solely on trying to remediate "incorrect" ideas held by students, as so much research has done in the past, it is important to try to reinforce correct ideas in situations in which we can identify them and then help students to place their correct ideas into an acceptable scientific framework. Concept substitution is a strategy that attempts to do this for a certain class of conceptual difficulties. This approach has a number of advantages. First, from an affective point of view, it is encouraging for students to hear that they are correct. Second, because students are not asked to give up their ideas completely, physics can make more sense than often happens in traditional approaches. Third, when the concept that has been substituted by the instructor is encountered later in the course, students already have some intuition about that concept. Fourth, the approach encourages students to make distinctions between related concepts that often remain undifferentiated in students' minds.

The positive results obtained after using concept substitution in the situations discussed in this chapter are very encouraging, particularly when one considers that the students concerned are both underprepared and are not first-language English speakers. Although the strategy has not been tested rigorously, it has potential as a means of promoting conceptual change and is worth exploring further.

REFERENCES

Clement, J., Brown, D. E., & Zietsman, A. (1989). Not all preconceptions are misconceptions: Finding "anchoring conceptions" for grounding instruction on students' intuitions. *International Journal of Science Education, 11*, 554–565.

Duit, R. (1993, January). *A constructivist view of learning science, especially physics.* Paper presented at the seminar "Curriculum Research: Separate Disciplines, Common Goals," Hilversum, The Netherlands.

Duit, R., Jung, W., & von Rhoneck, C. (Eds.). (1985). *Aspects of understanding electricity: Proceedings of an international workshop in Ludwigsburg 1984.* Kiel, Germany: Schmidt & Klaunig.

Erickson, G., & Tiberghien, A. (1985). Heat and temperature. In R. Driver, E. Guesne, & A. Tiberghien (Eds.), *Children's ideas in science* (pp. 52–84). Milton Keynes, England: Open University Press.

Hewson, P., & Thorley, R. (1989). The conditions of conceptual change in the classroom. *International Journal of Science Education, 11*, 541–553.

Jung, W. (1986). Alltagsvorstellungen und das Lernen von Physik und Chemie [Everyday ideas and learning of physics and chemistry]. *Naturwissenschaften im Unterricht—Physik/Chemie, 34*(13), 2–6.

Minstrell, J. M. (1984). Teaching for the development of ideas: Forces on moving objects. In C. W. Anderson (Ed.), *Observing science classrooms: Observing science perspectives from research and practice* (Association for the Education of Teachers of Science Yearbook, pp. 55–73). Columbus, OH: AETS.

Pfundt, H., & Duit, R. (1994). *Bibliography: Students' alternative frameworks and science education* (4th ed.). Kiel, Germany: Institute for Science Education (IPN), University of Kiel.

Physics Education Group (1988a). *Physics by inquiry: Electric circuits* (4th rev. ed.). Seattle: University of Washington.

Physics Education Group (1988b). *Physics by inquiry: Heat and temperature* (4th rev. ed.). Seattle: University of Washington.

Scott, P. H., Asoko, H. M., & Driver, R. H. (1992). Teaching for conceptual change: A review of strategies. In R. Duit, F. Goldberg, & H. Niedderer (Eds.), *Research in physics learning: Theoretical issues and empirical studies* (pp. 310–329). Kiel, Germany: Institute for Science Education (IPN), University of Kiel.

Shipstone, D. (1985). Electricity in simple circuits. In R. Driver, E. Guesne, & A. Tiberghien (Eds.), *Children's ideas in science* (pp. 33–51). Milton Keynes, England: Open University Press.

Vygotsky, L. S. (1978). *Mind in society: The development of higher psychological processes.* Cambridge, MA: Harvard University Press.

White, R. (1992). Implications of recent research on learning for curriculum and assessment. *Journal of Curriculum Studies, 24*, 153–164.

CHAPTER 14

Changing the Curriculum to Improve Student Understandings of Function

Jere Confrey and Helen M. Doerr
Cornell University, Ithaca, New York, United States

The significant revision of the existing mathematics curriculum that is called for in the United States by the National Council for Teachers of Mathematics (NCTM) (1989) includes an increased emphasis on contextual problems, multiple representations, and the use of computer technology. The *Curriculum and Evaluation Standards for School Mathematics* (NCTM, 1989) advocates that the concept of function has a central and organizing role around which many other important mathematical ideas revolve. As recognized in the research tradition on student conceptions, the success of these proposals depends crucially on the examination of student thinking about functions. Widely available new technologies that influence the course of learning challenge researchers to conduct this examination within rapidly changing contexts. Thus, the research must anticipate, critique, and influence these technological developments, while at the same time documenting and analyzing their implications for student understanding. In our research group, we have viewed this as a cyclic process of investigating student conceptions and developing innovative curricula. This has led us to a substantial reconceptualization of the function concept.

Research on student conceptions has matured over the last 15 years. In mathematics, early work focused on the documentation of student errors in the acquisition of formal knowledge (Confrey, 1990). More recently, attention has focused on bridging the gap between students' formal knowledge and practical understanding. This change signals the importance of moving from a failure-driven approach to the articulation of research-based curricula. To accomplish this, mathematics educators have elected to create multirepresentational environments and to use situations to create the need for functional concepts. Thus, students are provided with rich and open spaces for inquiry. Such curricula must integrate a developmental view of student conceptions and can benefit from the work of many science educators who argue for a series of models that lead toward increasing complexity in student understanding (Clement, 1993; Linn, 1994). Thus, in this chapter, we review briefly the issues surrounding student conceptions of functions, describe our research-based curriculum, and indicate the directions of our current investigations in modeling.

STUDENT CONCEPTIONS OF FUNCTIONS

In the conventional treatment of functions, a typical definition is "a function is a relation such that, for each element of the domain, there is exactly one element of the range." Such a definition does not necessarily preclude a multiplicity of approaches to functions. However, both in textbooks and in assessment measures, one sees a restricted view of functions marked by the dominance of an algebraic symbolism (Confrey & Smith, 1992). Graphs are seen as of secondary importance and tables a distant third. Furthermore, algebraically, the introduction of functions is nearly exclusively presented in $f(x)$ form. However, historic investigations have shown the critical role of tables, numeric calculations, and geometric investigations in the development of the function concept (Dennis & Confrey, 1993). Our research with students has shown that the rich diversity of multiple representations leads to a more robust and flexible understanding of functions.

Traditionally, functions are treated using a correspondence model that treats function as a mapping from x to a corresponding y. Our research has uncovered student uses of a covariational model that relates the change in x to the change in y. The students describe patterns of values within a single column, and they coordinate the values in two different columns to answer questions about the problem situation. For example, if one column increased additively by 2 and the other by 6, students would associate a change of 1 with a change of 3 and insert the appropriate values. Furthermore, this approach allows the students to describe situations in terms of rate of change easily. For instance, attention to rate of change is easily evoked by describing how x_1 changes to x_2 and y_1 changes to y_2, thus creating a relationship between x and y. Graphical representations and the use of first- and second-order differences in tables are also powerful contributors. This emphasis on the concept of rate of change and its complement of accumulation has facilitated functional thinking (Nemirovsky, 1994) and allowed for a gradual approach to calculus from the early grades.

The treatment of functions within multiple representations could ignore the question of how the function is generated and/or where its application is warranted. To avoid this, some (Monk, 1989; Roth, 1992) have argued for the value of contextual problems on the grounds that they are more socially relevant, realistic, open-ended, and inviting to students. In addition to these important qualities, we have chosen contextual problems to highlight the aspects of the function concept that are grounded in human actions. In the next section, we describe the curricular approach developed as a result of this research by describing the software, contextual problems, and the developmental approach to prototypical functions through transformations.

A CURRICULAR APPROACH

The software, Function Probe© (Confrey, 1991), was designed in response to our research on student methods for understanding functions. The software provides a rich set of tools for probing, linking, and coordinating multiple representations of those concepts. Function Probe© is familiar-looking enough to invite entry by practicing teachers and yet holds the potential to create intentional and dramatic change in the character of the inquiry in the mathematics classroom. Over the past five years, the research group has developed context-based curriculum materials (Confrey, 1992) to be used with the software to help create the kinds of changes envisioned.

Function Probe© encourages the portrayal of student actions through three representational windows: a table window, a graph window, and a calculator. Each window is designed to encourage the learner to operate actively in ways compatible within that representation. For example, in the table window, columns can be filled iteratively using simple operations or more complex relationships. Columns can be sorted and/or moved. Relationships between columns can be defined either through entering an equation or by linking two or more columns so that corresponding pairs of values always remain in the same row. New columns can be created either as differences or ratios of successive values in an existing column or as an accumulation of values in an existing column.

The graph window allows the student to enter directly discrete points or equations or to bring data in from the table window. The point sets or the equations can be translated, stretched, or reflected as functions. The "stretch" and the "translate" icons are designed to give the physical feeling of a stretch through a visual spring and a translation hand that coordinates with the physical movement of the mouse. Our approach to the design of this window sought to maintain the integrity and independence of graphical actions without undue restriction from algebraic representations. The transformational tool for stretching provides an illuminating example of the relative independence of representations. Because any stretch in the plane has a line of invariance, Function Probe© allows the student to place this line, called the "anchor line," in any vertical or horizontal position. We did *not* restrict the anchor line to be fixed at the axes for the convenience of the algebraic notation. Thus, for example, a student can move the vertical anchor line to a position at the vertex of a parabola and then stretch the parabola to a desired shape. The graph window also keeps a history of the actions taken by the student.

The calculator also keeps a keystroke record and allows the student to save a string of keystrokes by creating a new calculator button. Thus, the written record of calculator keystrokes becomes its own symbolic system that can be used to represent functional relationships. Students also can move functions and data among the representations. Emphasizing the independence of the representations has allowed us to reconceptualize how actions across multiple representations can be coordinated. For example, whereas most graphical software is algebra driven—changes in graphs can be made only by changing the algebraic parameters—Function Probe© allows a student to transform a graph directly through mouse actions. This approach thus allows the student to explore the relationship between visual changes in the graph and numeric changes in the discrete representation of the table. The symbolism of algebra then becomes one way of describing those changes. The goal in designing Function Probe© was to create a flexible tool that would allow and encourage students to explore mathematical situations in each of these representational forms.

CONTEXTUAL PROBLEMS

Piagetian research has stressed the importance of the evolution of human schemes through actions and the operations one carries out on those actions. Accordingly, we stress the development of operational schemes for understanding functions. For instance, to introduce linear functions, we pose the problem of laying out a stone path from a house to a bird feeder 44 feet away. There are to be 15 circular stones, each one foot wide, all equally spaced with the last one placed against the bird feeder. The first stone can be any distance from the house. Here, the invariance in the operation lies in the addition of each stone to the path. This recognition of the invariance of operations suggests the location of a function within a family of functions, or prototypic functions.

The prototypic functions that have been examined by the research group include linear (including absolute value and step functions), inverse, quadratic, exponential, and trigonometric. Algebraically, these are $f(x) = x$, $f(x) = 1/x$, $f(x) = x^2$, $f(x) = a^x$, and $f(x) = \sin x$, and so on. Each prototypic function, in some sense, is the simplest member of the family of functions and captures the basic action underlying a functional situation. Thus, the action of constant addition leads to linear functions; the action of splitting leads to the family of exponential functions; and constant cyclic action leads to trigonometric functions. Building an understanding of these repeated actions in contextual situations forms the basis of deeper understanding of the relationship between context and mathematical form.

To connect a prototypic function to characteristic operations and actions, we use contextual problems designed to help students to create and identify appropriate actions. It is the invariance of the operations and the actions that are carried out that locate a particular functional situation in its prototypical family. Furthermore, we believe that contextual problems should be central in the curriculum and should motivate learning in new topic areas, rather than follow formal mathematical presentations.

A flexible set of transformational tools provides students with multiple means of matching the prototypic function to the specific numerical measurements and parameters of a situation. Thus, for example, the basic cyclic action of a turning Ferris wheel might lead one to begin to model the distance between a seat and the ground with the prototypic function $d = \sin(t)$. A 40-foot diameter Ferris wheel making one revolution every 12 seconds could, for example, lead to a transformed function, $d = 40 \sin(\pi t/6) + 20$, through visual transformations in the graph window. The importance of contextual problems for multiple representations is that students are encouraged to seek out how the actions, operations, and roles are made visible (more or less) in the different representations.

MULTIPLE REPRESENTATIONS AND TRANSFORMATIONS

The development of multirepresentational approaches to function through contextual problems could appear to make the study of each prototypic function overly independent. Functional transformations are an important means of unifying this study of functions. Algebraically, the linear transformations can be expressed as $y = A\, f(Bx + C) + D$. That is, a linear transformation on the variable x is followed by a linear transformation on $f(x)$. Students learn that although all of the prototypes behave quite similarly under these transformations, the uniformity of that behavior is not necessarily obvious.

Transformations typically are introduced through vertical and horizontal stretches and translations of the identity function $y = x$, creating the class of linear functions. For example, in teaching linear functions, students need to build an understanding of how an initial amount and a constant additive rate of change can be seen as a y-intercept and the slope of a straight line, or as an initial value and an additive counter in a table. If students can come to recognize that these characteristics are necessary for a linear function, they are likely to be successful in solving contextual problems with linear functions.

Because of the equivalence of vertical and horizontal translations on lines, we use the absolute value function to introduce the distinction between a vertical and a horizontal translation. For example, whereas the graph of $y = (x - 2) + 6$ is identical to $y = x + 4$, it is not the case that the graph of $y = |x - 2| + 6$ is the same as either $y = |x + 4|$ or $y = |x| + 4$. The distinction between a horizontal and a vertical stretch is difficult to see in the absolute value function, but it becomes much more apparent in the step function $y = [x]$, particularly when introduced in relation to an appropriate context. We have used a parking garage fee structure as an example, showing that horizontal stretches (which affect unit time intervals) are clearly distinct from vertical stretches (which affect costs per unit time). This context allows one to explore separately the effects of each parameter in $y = A f(Bx + C) + D$.

Approaching functions visually is very compelling and satisfying for many students. Visual curve matching appears to depend on the graphs involved and it highlights the use of the stretch icon and the movable anchor line. For instance, when asked to match two parabolas, most students use a translation to match the vertices. They then proceed to stretch the functions to match the steepness of the curve. Students learn to move the anchor line to the vertex and to stretch away from the line (Afamasaga-Fuata'i, 1991).

One particularly visual use of the movable anchor line was demonstrated by a novice user at a teacher education workshop. In solving the stone path problem, this person imagined that all the stones could be bunched up against the feeder, going from 30 to 44, then a hook could be attached to the nearest stone and would be stretched, keeping equal spacing, until that first stone reached the house. On Function Probe©, she modeled the family of all possible solutions by using a vertical stretch (see Figure 14.1) of her first solution, placing the anchor line at 44 and then stretching until the first stone moved to the point (1,1), which would be up against the house.

Students typically understand translations relatively easily, but understanding stretching is more complicated. As they learn to use the stretch tool, they begin to understand what it means to hold a line of the plane invariant. However, they puzzle over the operational meaning of the stretching. To address this, Function Probe© has a resource for sampling the points on a curve. These points then can be sent to the table window where the impact of graphical actions can be examined and coordinated with the change in the table values. Predicting the algebraic outcomes that are a result of a graphic action and the graph outcomes from an algebraic action becomes a significant and multidimensional undertaking. For example, in Figure 14.2, the students create the initial graph, $y = \text{abs}(x - 2)$, and sample 15 points that are sent to the table. The students then are asked to explore stretching transformations and to develop

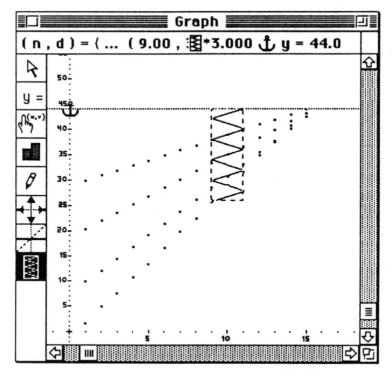

Figure 14.1. A Family of Solutions to the Stone Path Problem

x	y	(x)	(y)
original x	original y	stretched	stretched
-5.0	7.0	-2.5	7.0
-4.0	6.0	-2.0	6.0
-3.0	5.0	-1.5	5.0
-2.0	4.0	-1.0	4.0
-1.0	3.0	-0.5	3.0
0.0	2.0	0.0	2.0
1.0	1.0	0.5	1.0
2.0	0.0	1.0	0.0
3.0	1.0	1.5	1.0
4.0	2.0	2.0	2.0
5.0	3.0	2.5	3.0
6.0	4.0	3.0	4.0
7.0	5.0	3.5	5.0
8.0	6.0	4.0	6.0

Figure 14.2. Coordinating Graph Actions and Table Values

a relationship between each graph action and its effect on the numeric values in the table. In the case illustrated, the graph is stretched horizontally toward the y-axis by a factor of 0.5 with the anchor line at $x = 0$. The stretched data points then are sent to the table. The result of the stretch is that the x-values are cut in half and that y-values are unchanged. Exploring transformations with a set of points sampled from the graph, in combination with the table, can contribute dramatically to students' insights in this area and provide an example of how interactions among three of the representations can be woven together. For a more thorough discussion of the role of the anchor line and the use of transformed discrete points, see Borba (1993).

AN APPROACH TO MODELING

As an interactive software tool designed to foster the investigation of functional relationships, Function Probe© is an ideal tool for encouraging the expression of student-generated relationships among variables and for providing a flexible and powerful environment for investigating alternative relationships and possible conjectures. We are extending the use of Function Probe© into an integrated modeling approach that begins with physical experimentation to engage students in gathering data and posing conjectures. The students then move to the use of a simulation environment to explore conjectures, and the iterative process of developing and validating a solution through the use of multirepresentational analytic tools.

A curricular unit was designed to investigate motion down an inclined plane by integrating three components: the gathering of data from a physical experiment; the development and exploration of a computer simulation; and the mathematical (algebraic, graphical, and tabular) analysis of the data. What we have seen through the students' activities is a progressive process that simultaneously builds both a practical and a theoretical understanding of the behavior of an object on an inclined plane (Doerr, 1994). Rather than starting with physical laws expressed as abstract principles to be applied to contextual problems, this integrated modeling process simultaneously intertwines construction of the appropriate laws as they are needed, integration with existing student knowledge, and solution of contextual problems.

CONCLUSION

The approach to teaching functions discussed in this chapter was built with the recognition that students need to develop a concept of function that includes both a covariation and a correspondence notion of function. The instruction is grounded in human actions and operations that are tied to the development of families of functions through prototypic actions. We encourage the use and exploration of multiple representations, while striving to recognize the integrity, strength, and particular perspective that is illuminated by each representation in its own right. We have found that students build flexible and powerful understandings as they coordinate their grasp of actions in one representation with its impact in another. The approach to prototypic functions is integrated through the common framework of transformations within and across representations. Finally, we are extending the use of a multirepresentational analytic tool into a mathematical modeling approach that integrates experiment, simulation, and analysis.

REFERENCES

Afamasaga-Fuata'i, K. (1991). *Students' strategies for solving contextual problems on quadratic functions.* Unpublished doctoral dissertation. Cornell University, Ithaca, NY.

Borba, M. (1993). *Students' understanding of transformations of functions using multi-representational software*. Unpublished doctoral dissertation. Cornell University, Ithaca, NY.

Clement, J. (1993). *Types of analogies used in model construction*. Unpublished paper, University of Massachusetts, Amherst, MA.

Confrey, J. (1990). A review of the research on student conceptions in mathematics, science and programming. *Review of Research in Education, 16*, 3–56.

Confrey, J. (1991). *Function Probe* [Computer program]. Santa Barbara, CA: Intellimation Library for the Macintosh.

Confrey, J. (1992). *Learning about functions through problem-solving*. Ithaca, NY: Cornell University.

Confrey, J., & Smith, E. (1992). Revised accounts of the function concept using multi-representational software, contextual problems and student paths. In W. Geeslin & K. Graham (Eds.), *Proceedings of the Sixteenth PME Conference* (Vol. 1, pp. 153–160). Durham: University of New Hampshire.

Dennis, D., & Confrey, J. (1993). *The creation of binomial series: A study of the methods and epistemology of Wallis, Newton and Euler*. Unpublished paper, Cornell University, Ithaca, NY.

Doerr, H. M. (1994). *A model building approach to constructing student understandings of force, motion and vectors*. Unpublished doctoral dissertation, Cornell University, Ithaca, NY.

Linn, M. C. (1994). *Establishing a research agenda for science education: Project 2061*. Unpublished paper, School of Education, University of California, Berkeley, CA.

Monk, S. (1989). *Student understanding of function as a foundation for calculus curriculum development*. Paper presented at the annual meeting of the American Educational Research Association, San Francisco, CA.

National Council for Teachers of Mathematics (NCTM). (1989). *Curriculum and evaluation standards for school mathematics*. Reston, VA: Author.

Nemirovsky, R. (1994). Slope, steepness and school math. In J. P. da Ponte & J. F. Matos (Eds.), *Proceedings of the Eighteenth International Conference for the Psychology of Mathematics Education* (Vol. 3, pp. 344–351). Lisbon, Portugal: University of Lisbon.

Roth, W-M. (1992). Bridging the gap between school and real life: Toward an integration of science, mathematics and technology in the context of authentic practice. *School Science and Mathematics, 92*, 307–317.

PART III

Implementing Teacher Change in Science and Mathematics

CHAPTER 15

Analytical and Holistic Approaches to Research on Teacher Education

Kenneth Tobin
Florida State University, Tallahassee, United States

In the present era of curriculum reform, there has been a resurgence of interest in policy as it relates to teacher change; as an area of inquiry, however, teacher change may have suffered from too close a focus on classroom practices. Too often the bottom line in teacher-change research is evidence of altered teacher and student practices as a curriculum is implemented. Although the significance of obtaining such data is not questioned, lack of attention to the way in which teachers make sense of concepts *they* regard as critical to teaching and learning could have limited the scope of research questions, answers, and implications for the education of prospective and practicing teachers. In our efforts to ascertain whether reforms are being implemented as intended, researchers and policymakers could have placed too much attention on changes in the classroom without first looking to see whether conditions regarded as antecedent have been met. For example, if teachers are to change what they do while implementing science and mathematics curricula, it seems reasonable to assume that first they must want to change. Accordingly, researchers might investigate what teachers were doing prior to their endeavors to change and ascertain why they were acting in that manner. Then the inquiry could shift to teacher and student beliefs about the changes intended by would-be reformers.

The purpose of this chapter is to review what is known about teacher education from two points of view: first, a review of research and development on teacher learning and change; and, second, a synthesis of what we have learned from our ongoing program of research on science teaching and science teacher education. Research that used "analytical" approaches is distinguished from studies employing "holistic" approaches.

Analytical approaches to research on teacher learning and change encompass descriptions of teaching by relating classroom practices to theoretical models comprised of verbal statements of belief about knowledge, control, and restraints. On the other hand, holistic approaches encompass personal experiences of schooling and knowledge of teaching embedded in images of practice. For example, teachers describe teaching roles in terms of metaphors and associated images, and they can use these descriptions as foundations on which to build understandings of teaching and learning and strategies for changing the curriculum.

RATIONALE

In the present chapter and other chapters in this book, learning is viewed from a constructivist perspective (e.g., Tobin & Tippins, 1993). Within this view, learning is a social process of making sense of experience in terms of extant knowledge. From this perspective, learning to teach science is accomplished best by direct experience of the teacher-learner in conjunction with opportunities to reflect critically on the experience and emergent problems. Among those who find constructivism to be a useful way of thinking about teacher learning and change, there is a consensus on the basic tenets (see, for example, Duit and Confrey, Chapter 7, this volume).

Fosnot (1992) described learning as "a self-regulatory process of struggling with the conflict between existing personal models of the world and discrepant new insights, constructing new representations and models of reality as a human meaning making venture, and further negotiating such meaning through social cooperative activity, discourse, and debate" (p. 1). Fosnot's view of constructivism goes beyond beliefs about knowledge to include power relationships as she examines the social organization that, in her view, would result from the application of constructivist views.

The analytic perspective is evident in descriptions of teaching in terms of beliefs about knowledge and knowing, beliefs about power, and beliefs about restraints. What teachers believe knowledge is, and how people come to know, influences how they make sense of their roles as teachers. Also associated with their roles as teachers are beliefs about the relative power of teachers and students in the contexts that arise in the classroom. Finally, beliefs about restraints, or reasons for not doing what teachers believe that they otherwise must do, have to be considered in any account of what is happening in classrooms. The analytical perspective, too, is reflected in the following excerpt from Hewson, Zeichner, Tabachnick, Blomker, and Toolin (1992):

> Teachers build conceptual structures in which they incorporate classroom events, instructional concepts, socially approved behaviours, and explanatory patterns. These structures include their rationale for teaching (maybe implicitly), their view of knowledge, learning and science, their disciplinary knowledge, and the ways in which they teach, along with detailed specific information on content, students, school procedures. . . . It is clear that teachers' knowledge, skills, and attitudes are very different in kind, they serve different purposes, and are not necessarily coherent. Thus, it is reasonable to infer that teachers have developed a variety of alternative conceptions of teaching science. (pp. 4–5)

It is possible that teaching practices are learned and negotiated at a young age. As students engage in classroom activities over 12 or 13 years of schooling, they are exposed to many teachers and classmates. As a consequence, there is a multitude of opportunities to learn about teacher and student roles. Although it is not an explicit focus of K–12 schools, the teacher education curriculum is intensive (i.e., inadvertently students learn to teach). Of course, for prospective teachers, the curriculum extends throughout the college years (often prior to any formal commitment to become a teacher). By the time prospective teachers register for their first teacher education course, they have well-grounded knowledge of teaching and learning etched into their minds.

Throughout the lives of teachers, they continue to learn about teaching and learning as they teach. Daily experiences, interpreted in terms of prior knowledge, lead to learning that influences classroom practices in many cases. This knowledge is largely intuitive because teachers infrequently engage in formal discussions about teaching and learning (Clark, 1988). The processes of learning to teach in contexts such as these rarely involve formal analyses of teaching or reflections on the components of teaching. On the contrary, learning to teach in these contexts is experiential and knowledge of teaching is embedded in images of practice. This holistic, informal way of learning to teach seems to be effective, particularly because of the extended duration of the experience that begins in the kindergarten and extends through the professional lives of teachers. This approach stands in stark contrast to the analytical approach adopted in most formal teacher education programs.

Three terms have central importance in this chapter. *Belief* is knowledge that is viable in that it enables individuals to meet their goals in specific circumstances. Beliefs are tied to the situations in which actions are contemplated. *Action* consists of a behavior that can be observed, beliefs associated with the behavior (including goals), and a context in which the behavior is thought to be appropriate. *Referent* is a guide for action, is context specific, and is an organizer of the beliefs and actions an individual deems relevant to the situations in which personal actions are to occur. For example, if metaphors and associated images and language are referents for a teacher's actions, the referents can constrain possible teacher and student actions.

Within a context of the reform of science and mathematics teaching and learning, this chapter examines research on teaching and teacher education in terms of two approaches, analytical and holistic, and bridges built to link the two approaches. The chapter contains a review of the research of others, a review of the research of my research group, and conclusions.

LEARNING FROM OTHERS

The first subsection below reviews studies that employed a predominantly analytical approach, while the second subsection is devoted to research that used a holistic approach.

Analytical Approaches

Fosnot (1992) considers that teacher education programs should provide opportunities for teachers' beliefs to be illuminated, discussed, and challenged. What Fosnot has in mind when she thinks of reform is not just tinkering with a system in need of fine tuning, but a paradigm shift in which teachers and students engage in profoundly different ways. Fosnot planned a program for prospective teachers to engage in "learning experiences which confront traditional beliefs, in experiences where they can study children and their meaning making, and in field experiences where they can experiment collaboratively" (p. 18). Fosnot's multitiered approach to teacher education consists of four levels, or tiers, that involve both practicing and prospective teachers, together with a fifth tier that involves novice teachers, now teaching in their own classrooms. The five tiers are as follows:

1. Practicing and prospective teachers participate in coinvestigations of the processes of their own learning.
2. Practicing and prospective teachers construct a pedagogy from analyses of children's thinking.

3. Practicing and prospective teachers undertake cooperative fieldwork.
4. Practicing and prospective teachers engage in reflective fieldwork, focused observation, and/or research on a significant pedagogical issue.
5. Novice teachers engage in an integrative field experience, during which a faculty member from the project visits them one day a week.

Fosnot emphasizes the autonomy of learners and the provision of experiences in which they set the agenda for learning. Although the goals are negotiated, the approach to teaching that will work seems to be left to the individual and can be tailored to individual strengths and preferences and to the contexts in which teaching is to occur. The vision of what teachers, students, and classes ought to look like can be set by the teacher. In this respect, there is a difference between Fosnot's approach and that of Hewson and his colleagues, whose vision of what classes ought to be like is constrained by Hewson's conceptual change model (see Chapter 11, this volume), which assumes a preferred way to teach and that teachers should learn to teach in this way.

A key component of the approach advocated by Hewson and colleagues (1992) is action research, a form of collaborative self-reflective inquiry undertaken by teachers in order to improve their practices, their understandings of those practices, and the situations in which those practices are carried out. Teachers who engage in action research are more likely to self-monitor their teaching and become more aware of what happens in their classrooms, of the gaps between what they do and what they would prefer to do, and of what their pupils are thinking, feeling, and learning. The value of action research, for Hewson and colleagues, is that it enables teachers to view teaching as a form of inquiry or experimentation and thereby it encourages them to make conventional teacher and student roles problematic. As action researchers, teachers become learners in their own classrooms and tend to view their own experience as a source for learning.

Three of the chapters in this section follow this analytical approach. Baird and White (Chapter 16) describe some of their research into reflection, metacognition, and classroom change structures according to three guiding principles regarding the improvement of classroom teaching and learning. The following chapter (Chapter 17) by Northfield, Gunstone, and Erickson describes the assumptions and principles that underlie attempts to reform teacher education from a constructivist perspective. The final chapter of this section (Chapter 19), by Gallagher, describes how the teaching and learning in middle-school science and mathematics was transformed by the Support Teacher Program with its clearly defined goals and directions.

Holistic Approaches

In concluding that earlier approaches to conceptual change in mathematics teacher education were unsuccessful because they failed to change images, Wubbels, Korthagen, and Dolk (1992) advocated two alternatives: to reflect on images and metaphors in terms of analysis, argument, and traditional left-brain language functions, or to address images directly through the use of strategies usually associated with the right brain. For example, for a particular teacher, the metaphor of "teacher as lion tamer," evoked with words and/or a picture, might be associated with a range of classroom images. The generation of alternative images through the use of metaphor, sketches, drawings, photographs, and video vignettes appeals as a powerful technique to enable teachers to reconceptualize teaching and learning roles in holistic ways.

Munby and Russell (1990) suggest that it is productive for all teachers to become students of their own metaphors because how one describes the world gives insights into how one constructs it. Accordingly, the language of teachers provides an image of the professional knowledge of teachers and ways in which that knowledge might be represented and constrained. Numerous authors have conceptualized teacher and learner roles in terms of metaphors. Philion (1990) described the metaphors used by three student teachers: teacher as classroom manager, as referee and canvasser, and as a nurturer who was able to talk with students rather than to them. Roth (1991) described teaching and learning roles in terms of cognitive apprentice, scaffolding, and coach. Roth was not suggesting that every aspect of each of the three metaphors can be used to form an understanding of teaching and learning. Rather, meaning is associated with dynamic images such as a young learner apprentice with a wise master teacher, a sturdy scaffold reaching into new territory but firmly attached to extant knowledge, and a thoughtful diagnostician contemplating how to get improved performances from a hard-working athlete.

In a similar holistic manner, the chapter by Hand (Chapter 18, this volume) describes a model for in-service teacher education involving the diagnosis of the application of constructivist principles. The evidence for a particular teacher's achieving different levels of implementation of constructivist teaching and learning approaches was diagnosed in terms of the metaphors that teachers used to describe their teaching and an analysis of their teaching roles in terms of manager, technician, facilitator, or empowerer.

However, the use of metaphors as a focus for reflection is not useful always, according to Bullough (1991), because in some instances the construction of new metaphors does not represent a thoughtful reexamination of teaching and learning. In Bullough's research, the metaphors that emerged ranged from teacher as husband, teacher as butterfly, and teacher as policewoman to complex metaphors such as teacher as chameleon ("always changing and trying to find a spot where I am most comfortable," p. 49). The label (i.e., husband, chameleon) was not the characteristic that made the metaphor simple or complex, but rather the explanation of how teaching was related to being a husband or a chameleon.

LEARNING FROM OUR RESEARCH ON TEACHING AND TEACHER EDUCATION

Our program of research also can be conceptualized in terms of both analytical and holistic studies.

Analytical Studies

One of the most significant findings of our research has been the importance of particular cognitive referents as teachers decide to act in specific ways as the curriculum is implemented. Three referents that have been most significant are the teacher's *personal epistemology, beliefs about control*, and *beliefs about restraints*. The following examples involving Karl, a middle-school science teacher, illustrate how these three sets of beliefs relate to teaching and changing classroom practices. Karl developed a set of cognitive tools that could be used to think analytically about the way in which he taught and how he would like to change his teaching.

Tobin, Tippins, and Hook (1994) explained how Karl taught in a manner that was consistent with a personal epistemology of objectivism. However, he rarely thought about the nature of knowledge, nor did he use his beliefs about knowledge as a referent for action. Objectivism was a *default* epistemology embedded in his actions (also see Ernest, 1992). For example, Karl believed science to be a body of knowledge and truths to be learned by students, and he focused on content coverage and setting up activities to enable students to rote-learn the content he regarded as most significant. Covering the bulging textbook was a major goal of Karl's traditional approach to teaching. However, there were aspects of Karl's teaching with which he was dissatisfied and he had a commitment to personal change.

Karl's classes changed in numerous ways in the six-year period of our research. Like several other teachers in the school, Karl adopted a curriculum that had a heavy orientation toward fieldwork and problem solving. Although he learned about constructivism, he did not use it as a referent for some time, preferring instead to provide greater autonomy to students to enable them to make decisions with respect to their learning. Numerous ethical issues arose with respect to the extent to which Karl adequately supervised students and maintained a safe environment. At about that time, it became evident that Karl had worked out a new view of knowledge, one that was much more constructivist in character. Constructivism allowed Karl to counter the viability of a claim that he should be covering content. He became more comfortable with a focus on students' learning with understanding. Significantly, his beliefs about knowledge and students' having autonomy were integrated into a powerful framework used to make curricular decisions. Having reached this stage in his thinking, Karl began to build metaphors and images based on his beliefs about knowledge, autonomy, and constraints.

A study of Diane (Tobin & Imwold, 1993), a middle-school mathematics teacher, highlights the role of restraints in maintaining traditional practices. Diane believed there was more content than could be covered, or learned with understanding, in the time allowed. This referent is built on a belief that others had control over what was to be covered and learned in the curriculum. The power to decide what was best for her students did not reside with Diane, but with others such as state and district personnel, parents, school administrators, and colleagues. The restraint of time seemed to reflect Diane's feeling of disempowerment (i.e., an inability to determine her priorities and act on them).

In our research, Diane alluded to a potential problem with her colleagues, as well as with administrators, students, and parents. Consequently, she did not feel confident to introduce innovative teaching strategies if she could not count on support from her colleagues and others in the school. Diane viewed some students as posing problems by virtue of their attitudes, particularly when they were placed in groups for collaborative learning activities. Further, Diane believed that parents are a conservative force with respect to change. Because the school drew from regions of high socioeconomic status, it is not surprising that parents would be involved and provide input to the curriculum. From Diane's point of view, this was a deterrent to her desire to be innovative: The parent interest became a force that made her lean at times toward traditional practices. In general, parents are supportive of change but, when it comes to their children, tried and tested procedures must prevail. Risks could be taken with other children, not their own. This belief was reinforced by her experiences during the past years and was compounded by what she regarded as a lack of support from the administration for her novel approaches to teaching mathematics.

Our story of Diane shows a dialectical relationship between prior beliefs based on objectivism and control and emerging beliefs based on constructivism and student autonomy. She described herself as being in a transitionary period in which she knew in some ways what to do but, in other ways, she lacked the confidence to change from traditional practices and to confront the restraints that surrounded her.

Holistic Studies

An intervention occurred in a study involving Sarah, a middle- and high school science teacher, who was having a difficult time controlling her students (Tobin & Ulerick, 1992). Sarah thought about teaching in terms of the three roles of facilitating learning, managing students, and evaluating student performance. For each role, she used one or more metaphors to make sense of teaching and learning. The most critical role was management because the students were extremely disruptive. For this role, Sarah used a metaphor of teacher as comedian. Sarah believed that, if she used humor, students would like her and be less difficult to manage. With the help of a group of researchers, Sarah constructed a new metaphor for management. While thinking about teaching and learning from a constructivist point of view, Sarah considered roles outside of teaching that might be used metaphorically as a basis for teaching. Sarah was reconceptualizing an important teaching role in a way that was consistent with constructivism and that offered the possibility of improving learning. The metaphor she chose was the teacher as social director. She argued that a social director can only invite guests to a party. If they opt not to come, then the hostess can only make the invitation more attractive. If guests choose to come, however, they should come to have fun (learn) and ought not disrupt the fun (learning) of others. In addition, they should be courteous to the hostess and everyone at the party. Sarah readily related the metaphor to her role as a manager of students and used the metaphor to guide the implementation of new teaching strategies. In this way, many teaching strategies changed and student actions were constrained by the different actions of the teacher.

Almost immediately changes were evident. Sarah simplified a cumbersome rule structure so that only two rules would be enforced. All students were to respect the teacher and one another, and students were to come to class prepared to work. Five students who had been particularly uncooperative were assigned to another class until they could demonstrate that they could work without being disruptive. These changes gave Sarah the chance that she needed and, with relative calm in the classroom, she was able to implement additional changes to improve the quality of learning.

Sarah was aware that she conceptualized many of her teaching roles, such as her role as an assessor of student learning, in terms of metaphors such as the teacher as a fair judge. The metaphor of judge, and its associated images and language, guided Sarah's actions in all of the situations in which she assessed student learning. As such, the metaphor acted as a referent for teacher and student actions. The metaphor placed Sarah in the role of making judgments about the quality of student learning and she exercised considerable power over her students. Sarah decided the extent to which the knowledge of students was consistent with science as it is known and accepted by scientists. The "scales of justice" were Sarah's tools for making decisions. Either specific student knowledge measured up, in which case a student passed, or it did not, in which case the student failed. Much of what Sarah did was enacted as routines and was consistent with the fair-judge metaphor. On other occasions, when it was unclear what to do, Sarah consciously thought about herself as a judge and decided on an appropriate course of action. In this way, the referent constrained a set of associated actions of both Sarah and her students.

Sarah was dissatisfied with her assessment practices and the way in which they were perceived by students. Accordingly, she reflected on the fair-judge metaphor in relation to other constructs to which she attached value. Sarah decided to teach students in ways that were consistent with constructivism. Sarah's beliefs about constructivism became referents for her reflections about the appropriateness of the fair-judge metaphor. In the process of reflection, Sarah related possible actions of students and the teacher to the fair-judge metaphor and, in turn, that was related to her beliefs about knowledge and learning. When Sarah was unable to identify a suitable alternative, she asked the research team to make suggestions for her to consider. Each alternative was reviewed in relation to constructivism and possible classroom actions. Finally, Sarah declared that assessment was like a window into the minds of students, an opportunity for them to display their knowledge. At that time, Sarah made a commitment to reframe her assessment role in terms of a new set of guiding beliefs or a metaphor that, in her mind, was consistent with constructivism. When Sarah taught in accordance with her new referent, she was deliberative, referring each personal action and associated student actions to the metaphor. The use of the metaphor initially was conscious and overt as new routines were built to replace those that had been honed to be consistent with the fair-judge metaphor. In the process of implementing a new set of strategies, Sarah took account of her previous nonconstructivist approach to evaluation and informed all students that they could "start again" with respect to their evaluation. Students who previously were failing, and had no hope of elevating their science grades to the level needed to pass, now had a chance to get an A. Students seemed more committed to learning and their previously negative attitudes toward Sarah became more positive.

Over a period of more than a year, Sarah continued to reconceptualize her teaching in terms of new metaphors for additional roles (e.g., she added roles for being a teacher/researcher, a curriculum designer, and a staff development leader). In addition, she found that she was able to utilize metaphors that previously she was unable even to consider. The context in which Sarah taught had changed dramatically. Students now were prepared to construct themselves as learners and allowed Sarah to be their teacher.

The above examples concerning Sarah show how holistic representations of teaching and learning can serve as foci for reflection and analyses. When Sarah built a new metaphor, she did so by referring it to a set of beliefs associated with constructivism and control. This process is significant in that it links holistic and analytic re-presentations of knowledge of teaching and learning. One implication for teacher educators is that bridges can be built to link holistic and analytic re-presentations by focusing reflections in such a way that teachers identify the referents (i.e., sets of beliefs) underpinning given metaphors and images.

Three parts of a metaphor can be distinguished: a verbal part, an image, and a set of contexts in which the metaphor is regarded as viable. In terms of teaching, our research suggests that teachers can and do use metaphors to reflect on their teaching and to characterize roles they consider salient. Metaphors can be used as referents to constrain teacher actions while implementing the curriculum. Accordingly, changing metaphors appeals as a way of making significant numbers of changes in classrooms. First, Sarah could build her own metaphors to replace a metaphor that she felt no longer was appropriate as a referent. Second, when she could not think of a metaphor to replace her fair-judge metaphor for assessment, she accepted a suggestion by others and adapted a metaphor to her own emerging beliefs about teaching and learning.

This case study suggests that theory and practice can be linked by bringing together knowledge based on experience and holistic re-presentations, such as images and metaphors, with beliefs expressed as language, often associated with reading and discussions of teaching and learning. The power of the research is the linking of these ways of knowing about teaching and learning in the act of reflection.

CONCLUSION

The research reviewed in this chapter provides some insights into the challenges faced by would-be reformers. Perhaps a bridge between the analytical and holistic approaches to learning how to teach can be used as a basis for teacher education programs for prospective and practicing science teachers. Research on metaphor suggests that significant changes can be made to the implementation of a science or mathematics curriculum by "switching" metaphors (Hand, Chapter 18, this volume; Tobin, 1990). Reflection on professional actions can involve the identification of relevant metaphors and associated images and analyses of the associated referents and classroom practices. In this way, analytical and holistic approaches to teacher education are combined. A promising direction for research and development is to relate the use of images and metaphors to knowledge of teaching and learning that is accessible to language. The identification of additional ways of re-presenting knowledge of teaching and learning in a holistic manner is viewed as a priority. For example, if narrative accounts are regarded as holistic re-presentations of teacher knowledge, it is important to ascertain how they can be employed in the process of improving the quality of teaching and learning.

Holistic activities (e.g., involving the use of metaphor, images, and narrative accounts) can be used to describe and change teacher and learner roles. However, what is needed is a way to analyze holistic conceptualizations in terms of their embedded referents (e.g., in terms of beliefs about knowledge, control, and restraints) and to judge their efficacy. The use of analytical and holistic conceptualizations as tools for reflection might lead to conceptual change that is related closely to teacher and student knowledge-in-action.

The focus on the learning and change of individuals who want to initiate and sustain change is not sufficient to effect or to understand change. Because of the interactive aspects of cultures, it is necessary for all actors in a culture to undergo conceptual change if reforms are to be supported and sustained. Unless teachers and students receive adequate and continued support for their efforts to change, it is likely that the old-new dialectical struggle will lean toward the old and reform will be difficult to sustain. In this regard, it is useful to employ strategies within schools to support teachers engaging in change. Such strategies are described by the authors of the four chapters in this section. In addition, as was clearly evident in the case study involving Diane, it is not sufficient for teachers to have a commitment to personal change and a vision of what to do. Support is needed to identify restraints and overcome them and to empathize with those who succumb to the lure of tried and tested practices when new practices no longer seem viable. Understanding the implications of the old-new dialectic is essential if the reform process is to be sustained beyond initial, often abortive, attempts.

One way of building knowledge of teaching and learning is to involve teachers in research on their own classrooms. Action research, a component of most constructivist-oriented teacher education programs (e.g., Baird & White, Chapter 16, this volume; Hewson et al., 1992), provides a way of linking classroom practices and research, and allows teachers to express their goals for change. But what will be the foci for action research? How will teachers describe what is happening and frame questions about change?

The review undertaken in this chapter provides guidelines in terms of two somewhat different approaches to teacher education, each with the potential to address the continuing challenges of educating science teachers. Analytical approaches that relate classroom practices to theoretical models, comprised of verbal statements of belief about knowledge, control, and restraints, are effective in changing teacher beliefs and classroom practices. Research on conceptual change provides a heuristic for building approaches to science teacher education that can target identified practices for change, particularly in teachers' beliefs about teaching and learning. Such changes are necessary but insufficient conditions for reform. Other necessary conditions also must be met.

From the perspective of learning how to reform curricula, a productive focus for research is sociocultural analyses of teacher learning in relation to curriculum change. This has been a focus of this chapter. Although we have much to learn about teacher learning and curriculum change, we know a great deal that might inform efforts at reform. However, what is known and advocated frequently is ignored by policymakers who make sense from an objectivist perspective supported by beliefs about control that place the power to initiate and sustain reform at a great distance from teachers and local communities. Learning and subsequent change in policymakers and would-be reformers is a fascinating area of research and a critical focus area in dire need of attention.

In summary, this chapter reviewed research and development on teacher education from analytical and holistic perspectives, initially using a review of the relevant literature on teacher learning and change and, second, from a synthesis of an ongoing research program on science teaching and science teacher education. Consideration was given to analytical approaches such as conceptual change and an examination of teachers' personal epistemologies and beliefs about restraints and control. For the holistic approaches, consideration was given to helping teachers reflect on teaching roles in terms of such aspects as metaphors, images, cognitive apprenticeship, and scaffolding. By comparing the activities and outcomes of analytical and holistic approaches to teacher learning and teacher education, the chapter illustrates how both approaches can be combined to inform each other. The chapter concludes by recommending that both analytical and holistic approaches be used in appropriate ways by policymakers and would-be reformers.

REFERENCES

Bullough, R. V., Jr. (1991). Exploring personal teaching metaphors in preservice teacher education. *Journal of Teacher Education, 42*(1), 43–51.

Clark, C. M. (1988). Asking the right questions about teacher preparation: Contributions of research on teacher thinking. *Educational Researcher, 17*(2), 5–12.

Ernest, P. (1992, February). *The many are the one: Analysis and synthesis of papers on constructivism.* Paper presented at the State of the Art Conference on Alternative Epistemologies in Education, University of Georgia, Athens, GA.

Fosnot, C. T. (1992, April). *Learning to teach, teaching to learn: Center for constructivist teaching/Teacher preparation project.* Paper presented at the annual meeting of the American Educational Research Association, San Francisco, CA.

Hewson, P. W., Zeichner, K. M., Tabachnick, B. R., Blomker, K. B., & Toolin, R. (1992, April). *A conceptual change approach to science teacher education at the University of Wisconsin–Madison.* Paper presented at the annual meeting of the American Educational Research Association, San Francisco, CA.

Munby, H., & Russell, T. (1990). Metaphor as an instructional tool in encouraging student teacher reflection. *Theory into Practice, 29,* 116–121.

Philion, T. (1990). Metaphors from student teaching: Shaping our classroom conversations. *English Journal, 79*(7), 88–89.

Roth, W. M. (1991, April). *Aspects of cognitive apprenticeship in science teaching.* Paper presented at the annual meeting of the National Association for Research in Science Teaching, Lake Geneva, WI.

Tobin, K. (1990). Changing metaphors and beliefs: A master switch for teaching. *Theory into Practice, 29,* 122–127.

Tobin, K., & Imwold, D. (1993). The mediational role of constraints in the reform of mathematics curricula. In J. A. Malone & P. C. S. Taylor (Eds.), *Constructivist interpretations of teaching and learning mathematics* (pp. 15–34). Perth, Australia: Curtin University of Technology.

Tobin, K., & Tippins, D. J. (1993). Constructivism as a referent for teaching and learning. In K. Tobin (Ed.), *The practice of constructivism in science education* (pp. 3–21). Washington, DC: American Association for the Advancement of Science.

Tobin, K., Tippins, D. J., & Hook, K. S. (1994). Referents for changing a science curriculum: A case study of one teacher's change in beliefs. *Science & Education, 3*, 245–264.

Tobin, K., & Ulerick, S. (1992). An interpretation of high school science teaching based on metaphors and beliefs for specific roles. In E. W. Ross, J. W. Cornett, & G. McCutcheon (Eds.), *Teacher personal theorizing: Connecting curriculum practice, theory and research* (pp. 115–136). New York: Columbia University Press.

Wubbels, T., Korthagen, F., & Dolk, M. (1992, April). *Conceptual change approaches in teacher education: Cognition and action.* Paper presented at the annual meeting of the American Educational Research Association, San Francisco, CA.

CHAPTER 16

Metacognitive Strategies in the Classroom

John R. Baird and Richard T. White
Monash University, Melbourne, Australia

Educationists long have held the view that desirable learning involves understanding. It is interesting, then, to ponder why still so little about this process is understood. It is true that, increasingly, Australian school and educational research practices have started to reject the encyclopedic view of knowledge (associated with such passive, receptive metaphors of the learner's mind as *tabula rasa,* empty vessel, and sponge) in favor of a more active construction of understanding by the learner. However, the nature of this process of active construction remains far from clear. Two popular focuses of contemporary research into learning are *reflection* and *metacognition* (see Garner, 1990; Weinert & Kluwe, 1987, for a collection of essays on metacognition; White, 1988, for reviews). Even though the boundaries of association between these two concepts are ill-defined (some researchers even challenge the reality of the latter), research on them has served to clarify one aspect of activity during learning. This aspect is that learning with understanding is fostered when learners engage in *informed, purposeful* activity, to the extent that they exert adequate *control* over personal learning approach, progress, and outcomes. Indeed, these notions are central to the definition of metacognition itself: knowledge about, and awareness and control over, personal practice.

GUIDING PRINCIPLES FOR IMPROVING CLASSROOM TEACHING AND LEARNING

In this chapter, some of our research into reflection, metacognition, and classroom change is discussed. Three guiding principles emerged from the research, concerning who should change, the nature of the desired change, and how change should occur.

Principle 1. Improvement in classroom teaching and learning must involve change in students and teacher. For both students and teacher, change necessitates active processes of learning with understanding. Students must learn and understand more about the content, learning, and themselves; teachers must learn and understand more about learning, teaching, themselves, the students, and, quite possibly, the content.

Principle 2. Learning with understanding involves all of cognition (thoughts), affect (feelings), and behaviors. This deceptively bland statement has far-reaching implications for appropriate ways to go about improving teaching and learning. One implication, which will become increasingly apparent, is that neither this chapter nor any other can provide a set of "metacognitive strategies" that can be guaranteed to improve classroom practice. What we hope to provide, however, are some perspectives and some cautions that could guide attempts to foster classroom change involving attitudes, feelings, perceptions, conceptions, beliefs, and, accordingly, behaviors.

Our perspective regarding the nature of learning with understanding is central to our beliefs about how to structure change (considered below). We believe that learning with understanding derives from a process of *purposeful inquiry* leading to an outcome of adequate *metacognition*. Purposeful inquiry itself comprises two components that operate together: reflection and action. Reflection is *thinking and feeling:* asking such questions as "What am I doing?" and "Why am I doing it?"; selecting procedures to answer these questions; evaluating the results of applied procedures; and making decisions about what to do next. The second component, action, is *doing:* observing, manipulating, applying procedures, and enacting decisions. If carried out thoughtfully, this process of purposeful inquiry will generate a desirable level of metacognition; the person will know about effective learning strategies and requirements, and will be aware of, and be capable of exerting control over, the nature and progress of the current learning task. Baird (1992) provides a more complete discussion of the proposed relations between purposeful inquiry, reflection, and metacognition.

The brief description above underscores our meaning of a metacognitive strategy. The purpose of any metacognitive strategy is to generate information that will help people to be knowledgeable about, aware of, and in control of what they are doing. Metacognitive strategies are fundamental to purposeful inquiry; each strategy is initiated by asking oneself specific evaluative questions (reflecting) and implementing procedures to gain answers to these questions (acting). Next, we consider how metacognitive strategies can be used to improve classroom teaching and learning.

Principle 3. Change must provide for cognitive and affective growth. We start by reiterating the point that improvements in learning are as much related to changed feelings as they are to changed thinking. People like to feel informed and in control. They like to feel successful. The academic benefits that accrue from enhanced metacognition—improved performance and understanding—are matched by such affective benefits as greater satisfaction, fulfillment, and sense of purpose, control, and self-worth. In a classroom, this situation applies to both teacher and students. Teachers might initiate appropriate metacognitive strategies to evaluate and improve their teaching, as they endeavor to train students to apply analogous strategies in their learning. However desirable it might be, change is difficult. It often requires extensive personal development.

It is unlikely that strategies learned algorithmically and implemented mechanically will be truly metacognitive, in that they serve to enhance the learner's level of metacognition. What is needed is for the person undergoing change to be sufficiently intellectually and emotionally challenged by the activity to direct a process of purposeful inquiry. This person must be both *willing* and *able* to select and apply metacognitive strategies appropriate to the current purpose. Whether it be a student attempting to improve learning, or a teacher trying to improve teaching (to provide for better learning), we have found that four major conditions seem to be necessary in order to allow for the often considerable personal development entailed in directing purposeful inquiry. These conditions are *time, opportunity, guidance,* and *support.*

In the next two sections, some results from two major research studies on improving teaching and learning are reported. Even though these results are only a small sample of those obtained from over 10 years of research, they demonstrate the importance of each of the principles outlined above for effecting desirable change. More detailed information about these projects is available elsewhere (e.g., Baird, 1992; Baird, Fensham, Gunstone, & White, 1989, 1991; Baird & Mitchell, 1986; Baird & Northfield, 1992; White & Baird, 1991).

In comparison with much other research into teaching and learning, both research studies were unusual. They involved large numbers of secondary school students and their teachers in detailed collaborative reflection and action about everyday classroom practices. These teachers and students, in collaboration with university mentors, acted as researchers of teaching, learning, and change. For many of the participants, this reflection occurred over a period of months or years. The two projects were entitled Project for Enhancing Effective Learning (PEEL) and Teaching and Learning Science in Schools (TLSS).

PROJECT FOR ENHANCING EFFECTIVE LEARNING (PEEL)

This project commenced in 1985, at one secondary school in metropolitan Melbourne, and continues today in this school and about 20 others. The project was designed originally to extend the scope of earlier research by the first author into metacognitive training to improve the quality of classroom learning (Baird, 1986). In this earlier research, metacognitive training was based on eradicating seven "poor learning tendencies" (bad learning habits), each associated with inadequate metacognition, that had been identified. With the use of materials designed to assist them to ask reflective questions and determine answers, one teacher and three of his classes of students attempted to increase students' awareness of, responsibility for, and control over lesson activities. At the end of the six-month study, positive changes were recorded in students' attitudes and classroom behaviors.

The success of this earlier study led to PEEL, for which similar training was envisaged for many teachers and more students. In its first year, 10 teachers in seven subject areas and over 200 students from grades 7–10 participated. The method that was central to PEEL was Group Collaborative Action Research. There were two types of group collaboration. The first type comprised teachers and several university mentors attending a weekly out-of-class meeting to discuss progress, share information and experiences, and plan for the future. The second type was collaboration between each teacher and his or her students during class time. This collaboration involved teacher and students working together to make students' learning more active, informed, and purposeful.

Both types of collaboration produced interesting results, many of which were unexpected. Much of the teacher/mentor group collaboration centered on the idea of attempting to improve students' learning by replacing poor learning tendencies with more productive learning habits. At a full-day meeting during the second year of the project, the group approached this task by compiling a list of *good learning behaviors* that could be easily *observed*, and thus monitored, in the classroom. The group settled on a list of over 20 such behaviors, each of which exemplified active, purposeful inquiry. Examples of these behaviors were "actively attends," "checks work and corrects if necessary," "fits bits of work together/forms links," and "justifies opinion."

The group then compiled a list of 30 teaching/learning *procedures* that would foster such behaviors and that could be taught in the classroom. Some of these procedures had been developed elsewhere (e.g., concept mapping, Venn diagrams); others were new, being the product of the teachers' efforts over the preceding year. In subsequent years, PEEL groups in various schools continued to devise and adapt further procedures, such that the current list totals more than 80.

These procedures are not inherently metacognitive strategies. A procedure such as concept mapping can be used in many different ways, depending on the purpose for which it is used. The purpose is determined through *reflection;* a particular way of using concept mapping then might be employed in order to provide information that pertains to this purpose. For example, a student who is studying for a test and who wishes to answer the question "How much do I know about this area of work?" could choose to select some key terms and then attempt to arrange them into a map. Any terms that cannot be linked together convincingly then could be the focus for revision. The metacognitive strategy relates to the purpose of inquiry—here, diagnosing personal content understanding—and comprises the two components of inquiry, namely, reflection (asking the question) and action (employing the concept mapping procedure to find the answer). In another situation, with another purpose, the procedure of concept mapping again could be used, but perhaps in a different way, and even by a different person. For example, a teacher might decide to use concept mapping as a pedagogical tool to organize content for a new unit of work, to check students' prior intuitive beliefs in this content area, or to check personal teaching effectiveness by determining the nature and extent of students' understanding.

Some of the main points about metacognitive strategies can be summarized as follows:

- Metacognitive strategies are employed by a person engaged in a process of purposeful inquiry.
- These strategies comprise reflection (to determine purpose) and action (to generate information).
- Improved teaching or learning will follow improvement in the manner of implementation of appropriate strategies.
- Improvement is evidenced by enhanced metacognition (knowledge, awareness, and control of personal practice) and the associated affective benefits.

PEEL demonstrated the extent to which improvement in teaching or learning is difficult, demanding, and unsettling. The project started as an attempt to make students' learning more active, informed, and productive (thus the title Project for Enhancing Effective Learning). We envisaged that, in order to effect this improvement, students would need to undergo a process of substantial personal development. The teachers' job would be to determine how best to orchestrate this improvement.

What we had not realized at the beginning was the extent to which the project would place metacognitive demands on the teachers: In order that the project succeed, teachers had to undergo an intensive process of purposeful inquiry about their teaching. Most teachers found this process disturbing, as it involved detailed reflection on personal beliefs, perceptions, attitudes, and behaviors. Indeed, one of the most important findings from the project was that, for most teachers, substantial personal metacognitive development must precede metacognitive development in their students. For both teachers and students, the difficulties entailed in metacognitive improvement led us to identify the four conditions for change outlined earlier (namely, time, opportunity, guidance, and support).

Time

PEEL indicated that substantial and durable personal change requires more than days, weeks, or even months of effort. Any attempt at implementing school-based change must be designed so as to provide adequate time for personal and group change.

Opportunity

PEEL teachers stated strongly that one crucial benefit of the project was that it forced them to interrupt the interminable pressure of daily organization, lesson preparation, administration, and pastoral duties, and take time to reflect about themselves and their practice. Many teachers said that this was the first time since they had started teaching that such an opportunity had arisen and entered the project feeling dissatisfied with aspects of their teaching (particularly those aspects related to inappropriate student attitudes, behaviors, and performances). However, these teachers felt uncertain about what to do to improve the situation.

The metacognitive focus of the project provided an opportunity to grapple with some of these concerns in a deliberate, thoughtful manner. This focus encouraged teachers to become more reflective, and it required them to reconceptualize fundamental aspects of the nature and processes of teaching and learning, particularly their role in classroom activities.

In an analogous fashion, the project provided novel opportunities for students to take a more reflective approach to their classroom learning. In terms of subject content, students were unfamiliar with lessons in which they were given opportunities to ask reflective questions of themselves or others, and the time to seek answers. Similarly, students required opportunities to examine their ideas and beliefs regarding the nature of desirable learning, and to consider the demands and benefits of a more active, metacognitive approach.

Guidance

Initially intensively, but more intermittently over the years, the authors and some other university staff have participated in teacher group meetings in PEEL schools. The role of these university mentors has been to act as guides and critical friends, to encourage initiatives, to introduce ideas, and to place the teachers' findings into the wider educational context. One especially important role of these tertiary participants has been to emphasize the importance of having the teachers continually clarifying their *purpose*, by asking them such questions as "Why did you use that procedure?"; "What were you trying to achieve at that time?"; and "What do you think you should try next, and why?"

Similarly, students require clear guidance in attempting to become more active and informed learners. As this guidance often must come from the teacher, it follows that teacher metacognitive development must precede that of the students. Thus, it would be expected that PEEL—set up to improve students' learning—first would have to be a teacher development project. Such is the value of hindsight.

Support

For both teachers and students, a primary source of support in sustaining difficult and demanding development was the group collaboration. Sharing the endeavor with others proved a powerful incentive to continue. In a book produced to describe the first year's experiences of PEEL (Baird & Mitchell, 1986), teachers recounted the benefits of being able to share successes and, perhaps more importantly, failures with supportive "kindred spirits." Many of these teachers became increasingly committed to the rationale and procedures of PEEL, as they started to benefit from the assurance and satisfaction associated with an enhanced sense of awareness and control over their teaching. Increasingly, students were rewarded by PEEL, as teachers started to support and reward changed classroom attitudes and behaviors.

Mitchell (1992) identifies support with trust: The students must trust the teacher to value their opinions, to manage the classroom fairly, and to resolve confusion; and they must trust each other to be supportive. But these forms of support are within the classroom; forces from outside must also be in accord. Often they are not. Examination systems, forms of school organization, a general view in society that learning consists of "cramming," all could inhibit development of metacognition.

THE TEACHING AND LEARNING SCIENCE IN SCHOOLS (TLSS) PROJECT

This project also was based on the method of group collaborative action research, but it did not start with PEEL's overt metacognitive training objective. The aim of TLSS was to explore teachers' and students' views of what constitutes *quality* in science teaching and learning in classrooms, and to have teachers and students join together to research ways of improving this quality. The project ran for the four years 1987–1990. Over this period, it comprised 33 separate but related research studies; the longitudinal nature of the research meant that one study influenced the nature of the next (Baird, 1992; Baird et al., 1991). While the project was centered on secondary science teaching (involving more than 41 teachers and over 2,000 students from grades 7–11 at 15 schools), smaller numbers of teachers and students at elementary and higher education levels also were involved. Below, only findings from the secondary level are considered.

Again, there were two types of group collaborative research that involved teachers, students, and university mentors. These two types were regular out-of-class meetings (this time often attended by representatives of all three classes of participant) and within-class activities. For different participants, the nature and extent of collaboration varied. Some participants simply completed a questionnaire or were interviewed once, or, at the most, a few times. For 26 of the secondary teachers and over 500 seventh- to tenth-grade students, however, involvement was more intensive and long-term, lasting from several months to three years.

Two of the TLSS studies are considered below to illustrate features related to the three guiding principles given earlier. The first study involved out-of-class collaborative research by groups of teachers and mentors, and the second study involved classes of students, their teachers, and a mentor.

The first study occurred in the project's first year, 1987. Four teachers at one school and five teachers at another met weekly with one or more mentors to discuss research interests, determine a research focus, and plan research activities. Between these group meetings, a mentor observed each teacher several times in his or her class. After each of these lessons, teacher and mentor shared their perceptions and beliefs about what had happened during the lesson. The findings from these discussions were reported back to the teacher group. Based on these findings, the group then developed the research focus, which was to determine reasons why many secondary students become increasingly disenchanted with science as they progress from grade 7 through secondary school. Although details of the research carried out by these two teacher groups in 1987 are not given here, they are reported in Baird et al. (1989). However, the research done in that year served to highlight the key notion of *challenge,* which became the basis for all of the succeeding research. What we would like to emphasize here is the way in which this teacher group research illustrated each of the guiding principles above.

As with PEEL, this teacher group research demonstrated that, in order to improve classroom practices, teachers must undergo a process of cognitive and affective change. During 1987, these teachers experienced considerable change in their attitudes, beliefs, and conceptions regarding the nature of desirable classroom teaching and learning, and how to act in order to promote such teaching and learning. In order to change, however, they required each of the four conditions mentioned above. Teachers needed a period of many months in order to work through the ideas being shared. Perhaps surprisingly, much of this time was required to practice systematic, detailed reflection on teaching and learning—something most of them had not done previously. The opportunities provided by the study to attempt such reflection in a guided and supportive context proved to be among the outcomes most valued by these teachers. Overall, there was considerable evidence that, by the end of the year, most of the teachers were adopting a more reflective approach to their work. They more regularly applied metacognitive strategies to monitor their teaching and their students' learning, with the result that they felt more aware of and in control of their actions.

The second study also was based on a shared reflective process, this time involving students. Students were invited to collaborate in the investigation of what constitutes quality in science teaching and learning, and to work with the teacher to improve this quality. This study, entitled the "Agreement for Change," started with an extended class discussion between teacher, mentor, and students in an attempt to identify all of the features of the current science classes that diminished students' levels of application, understanding, and enjoyment. Major features, arranged in mutually agreed-on categories, were written on the chalkboard. Teacher and students then identified and agreed to three changes that the teacher would make and three changes that the students would make to classroom behaviors in order to ameliorate selected features. Examples of agreed-on teacher changes were providing more variety in lesson activities, giving clearer instructions, and using simpler language. Student changes included asking more questions, completing set work on time, and assisting each other more often. After entering into the agreement, teacher and students then attempted to institute the changes. The progress of the agreement was monitored by means of a form completed regularly by the teacher and each of the students. This procedure was undertaken by 12 teachers and 316 students in 14 classes over grades 8–11 at five schools. Overall, the procedure was very successful. Depending on the class, the agreement lasted up to 14 weeks. In sum, students remembered the agreed-on changes, many believed that their teacher was acting to make changes (resulting in an improvement in their enjoyment and understanding in lessons), and most believed that the changes that they themselves were making were leading to enhanced enjoyment, application, and, very importantly, understanding of the work (Baird et al., 1991). In this study, students and teacher had collaborated to form a working group based on shared needs, concerns, responsibilities, and accountability. The study exemplified each of the guiding principles above and, particularly, provided all of the conditions necessary for sustained change by both teacher and students. Based on the specific purposes they had set themselves, the nature and structure of the agreement provided for a systematic, purposeful approach to classroom actions through application of appropriate metacognitive strategies.

CONCLUSION

These projects have enabled us to determine useful procedures of teaching and to specify indicators of high quality in learning. The strong implications from our research are that time, opportunity, guidance, and support must be made available if students, and teachers, are to engage in purposeful inquiry and develop appropriate metacognitive strategies. We cannot, however, give a simple prescription to develop rapidly high-quality teaching and learning in any given classroom. That is because learning is complex, affected by a large number of factors. What happens in a classroom depends on the mix of personalities, the participants' earlier interactions with each other, their perceptions of their purpose for being there, and their physical and emotional states. It depends also on social forces from outside—the attitudes of parents, school authorities, and other teachers and students, as well as the actions of the government. Consequently, each classroom in unique, and what works in one may fail in another. The impossibility of a simple solution, however, is not a reason to despair. Despite the differences present between the many classrooms in our studies, we observed strong progress in the development of meta-cognitive strategies and consequently in the quality of learning.

REFERENCES

Baird, J. R. (1986). Improving learning through enhanced metacognition: A class-room study. *European Journal of Science Education, 8*, 263–282.

Baird, J. R. (Ed.). (1992). *Shared adventure: A view of quality teaching and learning* (Second report of the Teaching and Learning Science in Schools Project). Melbourne, Australia: Faculty of Education, Monash University.

Baird, J. R., Fensham, P. J., Gunstone, R. F., & White, R. T. (1989). *Teaching and learning science in schools: A report of research in progress.* Unpublished monograph, Monash University, Melbourne, Australia.

Baird, J. R., Fensham, P. J., Gunstone, R. F., & White, R. T. (1991). The importance of reflection in improving science teaching and learning. *Journal of Research in Science Teaching, 28*, 163–182.

Baird, J. R., & Mitchell, I. M. (Eds.). (1986). *Improving the quality of teaching and learning: An Australian case study—The PEEL project.* Melbourne, Australia: Monash University.

Baird, J. R., & Northfield, J. R. (Eds.). (1992). *Learning from the PEEL experience.* Melbourne, Australia: Faculty of Education, Monash University.

Garner, R. (1990). When children and adults do not use learning strategies: Toward a theory of settings. *Review of Educational Research, 60*, 517–529.

Mitchell, I. J. (1992). The class level. In J. R. Baird and J. R. Northfield (Eds.), *Learning from the PEEL experience* (pp. 61–104). Melbourne, Australia: Faculty of Education, Monash University.

Weinert, F., & Kluwe, R. (Eds.). (1987). *Metacognition, motivation and understanding.* Hillsdale, NJ: Erlbaum.

White, R. T. (1988). Metacognition. In J. P. Keeves (Ed.), *Educational research, methodology and measurement: An international handbook* (pp. 70–75). Oxford, England: Pergamon.

White, R. T., & Baird, J. R. (1991). Learning to think and thinking to learn. In J. B. Biggs (Ed.), *Teaching for learning: The view from cognitive psychology* (pp. 146–175). Melbourne, Australia: Australian Council for Educational Research.

CHAPTER 17

A Constructivist Perspective on Science Teacher Education

Jeff Northfield
Monash University, Melbourne, Australia

Richard Gunstone
Monash University, Melbourne, Australia

Gaalen Erickson
University of British Columbia, Vancouver, Canada

> Many students took what they were taught about the scientist's portrayal of the world only as a veneer of knowledge, beneath which they maintained the contradictory beliefs that they had formed earlier from unguided interpretations of experience.
>
> —R. T. White

The quotation above is a summary of findings from research on students' understanding of scientific concepts over the last two decades. This research has had a significant impact on curriculum, teaching, and learning in science. However, the impact on education has been superficial. Very often, the same observation made about students and their learning of science can be made about teachers and how they learn about teaching.

Preservice teachers bring at least 15 years of formal educational experience to their preparation for a teaching career. This extensive experience as a learner is powerful in shaping beliefs about teaching and learning, but is limited because it lacks the perspectives given by the teaching role and by consideration of alternative beliefs. The experience leads, inevitably, to a unique construction of schooling. Teaching requires a concern for a wide range of abilities and interests in students and a wider and more informed understanding of the teaching role. The extensive but limited previous experience of student teachers alone cannot give this.

Teacher education programs that involve examination and reconstruction of the prior beliefs and experiences of teachers have proven to
be a considerable challenge for teacher educators. Where constructivist
ideas have been applied to learning about teaching, fundamental changes
have been needed. However, these changes have been difficult to implement within the existing constraints imposed on teacher education programs.

This chapter begins by setting out the assumptions and principles
that underlie attempts to reform teacher education in terms of a constructivist perspective (i.e., the fundamental changes that we argue are indicated by constructivist ideas). This is followed by a brief outline of some
specific ways in which these principles have been applied in selected
teacher education programs. The final section addresses the implementation of such fundamental changes in teacher education.

APPLYING PRINCIPLES OF CONSTRUCTIVIST LEARNING
TO TEACHER EDUCATION

Preservice and in-service phases of teacher education are not distinguished here because the same principles apply throughout a teacher's
professional career. Indeed, the idea of career-long professional development would be enhanced by a coherent set of such principles. Four fundamental principles that can be derived from constructivist ideas are considered below and are linked with practice. Particular cases outlined in
the next section highlight the differing experiences that are necessary
when applying the principles to teachers who have different concerns at
different stages in their careers.

*Principle 1. Teacher education should strive to challenge existing concepts
and lead to conceptual change on the part of teachers.* This principle is not at
all controversial. Teacher educators, regardless of their desired outcomes,
value programs that generate some real change in the thinking and behavior of teacher education students. In order to see the relevance of the
principle to the arguments of this chapter, one must consider the principle in the specific context of constructivism; the next principle is the
result of this constructivist consideration.

*Principle 2. Teachers are learners who continually and actively are con-
structing their views of teaching and learning, which were based on personal experiences and which are shaped strongly by prior ideas and beliefs.* This principle
is fundamental to a constructivist view of learning. Table 17.1 lists some
key features relevant for those seeking to review their teacher education
programs in the light of constructivist learning ideas.

Table 17.1
Implications of a Constructivist Perspective
for a Teacher Education Program

Aspect	Activities Needed to Promote Professional Development
The initial perspective	Assess teachers' present strengths and existing values about and perceptions of teaching and learning
Planning the program	Utilize and build on existing teacher strengths, beliefs, perceptions—extending their present skills rather than undermining them; this begins by helping teachers to perceive their own existing strengths, beliefs, and perceptions
Learning process for teachers	Assess which aspects of the content and format of any new idea will be likely to appeal to the teachers and find ways of encouraging *reflection on practice*; present ideas so that they are *intelligible, plausible,* and *fruitful*
Impact of teachers	Be sensitive to the five possible outcomes of the introduction of ideas: • simply rejected • misinterpreted to fit in with, or even support, existing ideas • accepted, but the teacher cannot apply them in another context • accepted but lead to confusion • accepted and form part of a coherent long-term personal view of teaching and learning
Summary of a constructivist approach	Seek out feedback and provide experiences and follow-up activities to allow teachers to rethink and discuss their ideas

The phrase "reflection on practice" (see Table 17.1) has become used widely in teacher education, but too often it is interpreted in superficial ways. Schön's (1983, 1987, 1988) ideas have focused attention on the importance of experience in the process of conceptual change, but we would urge a closer examination of the conditions necessary to make experience a significant factor in learning (Erickson & MacKinnon, 1991).

The emphasis on presenting ideas that are "intelligible, plausible and fruitful" (Hewson, 1981) serves as a reminder that teacher education is about teacher learning and must be connected to the experiences and concerns of teachers. In a sense, these terms establish some criteria for developing learning experiences that can engage teachers and probe existing ideas and beliefs. In preservice teacher education, these three terms encapsulate the challenge of structuring a curriculum in a way that relates to a student teacher's limited experiences and concerns. At this stage of a teaching career, it is important to consider the three broad areas of learning and the prior experiences and beliefs that shape the learning outcomes:

1. learning about teaching and learning—the professional knowledge components of the program;
2. learning about subject matter, both to supplement academic backgrounds and to see the "content" from the perspective of someone who has to encourage others to learn; and
3. learning about themselves as they make the transition to being teachers—a challenging but often neglected component of teacher education programs.

The range of possible outcomes (see Table 17.1) highlights the way in which learning experiences interact with prior experiences and context.

Principle 3. The professional knowledge component of teacher education requires continual development and a primary focus on the classroom setting. This principle applies to preservice education (for which the role of the practicum is clearly critical for effective teacher education) and to in-service education for which some of the most promising approaches have involved teachers in researching their own classroom practice (Baird & Mitchell, 1986; Baird & Northfield, 1992; Erickson, 1991).

Principle 4. Learning about teaching requires support for teachers and conditions that allow them to interact with others about teaching and learning experiences. The interpretation of experience and establishing an identity as a teacher require interaction with peers. Professional development cannot happen to teachers on their own and it is important that student teachers are introduced to teaching as a collegial profession.

This set of four principles has emerged from a consideration of the implications of constructivist views of teaching and learning for teacher education. To apply these principles to teacher education programs would result in major reforms of existing practices. The next section outlines some examples of major changes in teacher education based on the principles set out above. In the final section, some reasons for the limited impact of such principles on teacher education programs are introduced.

THREE EXAMPLES OF THE APPLICATION OF CONSTRUCTIVIST IDEAS TO TEACHER EDUCATION

Collaborating with Teachers

The authors have been associated directly with two professional development initiatives designed to provide conditions for teachers to collaborate, to explore, and to respond appropriately to the growing research based on student ideas in science. In Canada, there is a group known as the Students' Intuitions and Science Instruction group (SI^2 Project) (Erickson, 1988), while in Australia the Children's Science Group is a network of teachers (Gunstone & Northfield, 1988). Each group consists of teachers and academics meeting to share and develop their understanding of science teacher education. In attempting to structure the group agenda and activities according to the principles outlined in the previous section, several unique features have emerged.

First, the leadership and agenda are very much in the hands of the classroom teachers. To some extent, teachers are using the group to gain ideas and check their growing understanding of science teacher education as they make personal constructions of the diagnostic and teaching strategies appropriate to their classroom settings.

Second, the original purposes and rationale for establishing each group have been reshaped considerably in the light of experience. Erickson (1988) commented: "We did not agree on a specific blueprint for action, nor was there much enthusiasm for developing detailed teaching strategies for the content areas that had been the subject of our earlier investigations into students' intuitive ideas" (p. 7). In both groups, there has been an emphasis on identifying and documenting more generalized teaching strategies and descriptions of practice that illustrate the constructivist perspective everyone is trying to understand. This has meant that much time has been spent striving for a common language to communicate the perspectives and tacit understandings the members of each group share. The shift in the rationale for this approach links with changing views of conducting research in the area of teaching and learning:

Briefly stated, my original pragmatic arguments were that teachers must be actively involved in the project because they had direct access to the classrooms where the teaching strategies were to be field tested, they were in a better position than university researchers to judge whether the strategies were effective and functional, and finally they would have more credibility with their peers in terms of communicating our results to other teachers. The theoretical argument was based primarily upon Schön's analysis of reflective practice, which leads to a position that the development of pedagogically functional knowledge must be constructed in the practice setting. (Erickson, 1988, pp. 14–15)

The members of each group learned about constructivist teaching and learning by practicing the principles in their own professional development. The groups also provided opportunities for the academic researchers to be learners about the nature of research and development in this area:

The academics in the group have consistently viewed the group as a meeting of equals, some of whom have professional knowledge and some of whom have more contextually isolated and more formalized research knowledge. The academics have undergone considerable conceptual change. (Gunstone & Northfield, 1988, p. 14)

This case example highlights the importance of providing conditions for teachers and academics to collaborate as they strive to understand constructivist ideas and their implications for classrooms. Such understanding requires personal constructions of meaning and, if the constructivist research effort is to have significant impact on schooling, it must lead first to reform in the way in which we conceive of and organize teacher education.

The PEEL Project

Because extensive accounts of this Project for Enhancing Effective Learning (PEEL) already exist (see Baird & Mitchell, 1986; Baird & Northfield, 1992), only a brief description is provided in this chapter. The project points the way to developing research from a constructivist perspective (e.g., White, 1990). Providing conditions for teachers to undertake action research in their classrooms also has been associated with powerful professional development effects on all participants. PEEL is therefore an example of a teacher education approach based on the principles outlined above and the view that learning about teaching should be considered in a similar way to learning in the classroom.

The professional development was in the hands of teachers, but with academics acting as resources and reactants. The school administration had structured the school timetable to enable the group of 10 teachers to meet each week. An informed and committed change agent (John Baird) linked the school with external resources, assisted in documenting the process, supported the teachers and students through a difficult process of change, and carried out research into the way in which constructivist and metacognitive ideas interact within school settings. The result has been a model of teacher education and a network of people to guide and support groups of teachers in other schools.

The PEEL project is an example of teachers' leading the way in research and development as they designed and refined teaching strategies and developed alternative approaches to assessment. Their findings have shaped subsequent curriculum reform in the state of Victoria in Australia in significant ways.

Toward Reform of Preservice Education

There would be few preservice programs that do not include some consideration of the constructivist perspective. However, applying the principles set out in the previous section would require a major restructuring of existing preservice programs; unfortunately, documentation of such reforms is rare in the literature. In most institutions, preservice teacher education is carried out by staff with a wide variety of responsibilities. Reforming preservice education would require the involvement of academic staff who have high status and who see the initial preparation of teachers as their primary function. It also would appear necessary for academic staff to have the commitment to reform that comes from research and development in constructivist ideas and their implementation.

One example of a reform of preservice teacher education has been documented to a limited extent (Gunstone, Slattery, & Baird, 1989; Northfield & Gunstone, 1983; Walker, 1989). Applying the principles set out earlier in this chapter to a preservice program required establishing a group of staff and student teachers to undertake a program separate from the mainstream program. This alternative approach had several important characteristics.

First, a major part of the preservice course was undertaken involving a small group of 15 student teachers working closely with two staff members. The collaboration so important to the success of the previous two case examples also was important in assisting new teachers to make a personal construction of their role as a teacher.

Second, the arrangement of teaching/learning experiences is critical as theory is assumed to emerge from considering the practice of teaching. Limited microteaching, involving one teacher with one pupil in teaching/learning sessions, and incidental teaching to enthusiastic lower elementary school groups are among the graded series of experiences that formed the basis of collaboration and "theory" development in the initial weeks of the program. In recent years, this has been followed by a 10-week period in a school when four to six members of the group experience the full range of teaching responsibilities, and seminars are conducted in the school setting to develop the theory as it emerges from experience.

Third, the professional and foundation studies ideally are integrated and used to clarify and develop the experience. This requires flexible scheduling and timetables that change from week to week, although experience over more than a decade has led to a more predictable program throughout the total course. Independence from institutional demands and timetables has allowed staff and student teachers to spend an extended period in a camp to prepare and conduct a field trip for a school population as a basis for discussion of field excursions.

Application of the principles of constructivism to preservice teacher education has been difficult to introduce and maintain. It has required staff to have such a commitment to preservice education that they will devote class time well beyond normal expectations for the purposes of providing and reviewing the experiences and activities developed as part of the program.

After more than a decade, the program still exists as an alternative approach taken with secondary science teachers within a larger mainstream program. Recent years have seen a shift toward constructivism in the mainstream program and the introduction of a graduate elementary teacher education program that recognizes constructivist ideas and the value of the activities associated with the original reform.

ISSUES IN APPLYING CONSTRUCTIVIST IDEAS TO TEACHER EDUCATION

In this chapter, it has been assumed that the constructivist research thrust has clear implications for teacher education. These implications have been developed in a series of principles, with three case examples outlined from the limited literature describing details of the impact on teacher education. The authors of this chapter claim some success at reforming teacher education in the light of constructivist principles, although the changes have been achieved with great difficulty. This final section therefore can be regarded as most important for readers who might wish to reform teacher education along the lines outlined in this chapter. Such reform will require that the issues discussed below be addressed during the process of change.

Limited Impact of the Constructivist Perspective

During 1988, each of the 50 institutions in Australia concerned with the preparation of teachers was visited as a part of an extensive review of Mathematics and Science Teacher Education (Department of Employment, Education and Training, 1989). One member of the review team observed that, while almost every institution had included constructivist developments in its teacher education programs, in almost every case it was a "one-off" lecture with perhaps some discussion. The evidence was clear that, although constructivism commonly has become part of the content, Australian teacher education programs largely have been unaffected by an *ideology* that has implications for the whole approach to learning to be a teacher.

If we extend our considerations to in-service education, the impact of constructivist ideas on teachers and teaching and learning also is limited. The first two cases in this chapter are rare examples of in-service work that has been documented. The PEEL books (Baird & Mitchell, 1986; Baird & Northfield, 1992) have created interest among teachers. Pedagogical exemplars (Erickson, 1988), in the form of videotapes to illustrate teaching/learning situations, also appear to be a promising line of development that allows teachers to see what might be possible in terms of putting initially abstract ideas into practice.

Some recent research (Baird & Northfield, 1990) indicates that in-service activities, while successful in the development of individual teachers, could be having minimal impact on teaching and learning in the school. This brings attention to the issues of "innovation focused" rather than "schoolwide focused" efforts to bring about change (Fullan, 1988). There is a challenge for teacher education (and school improvement efforts more generally) to develop approaches that can extend impact beyond the development of individual teachers and into wider school policies and practices. The collaboration so crucial for professional development of teachers must be extended more widely (e.g., Erickson, 1991) if constructivist ideas are to affect schooling in significant ways.

Creating Conditions for Teachers to Learn About Teaching

The two in-service cases outlined in this chapter involved teachers who volunteered to go far beyond the day-to-day demands of teaching.

Collaboration had to be developed among a group of teachers for whom isolation previously characterized their working conditions. The two cases can be considered as ways of creating conditions for learning that normally do not exist for teachers.

The difficulties associated with establishing and maintaining the teacher education approaches described in this chapter must be emphasized. They mirror the difficulties and issues with which all teachers will have to contend as they shift to a more constructivist environment for learning. Such difficulties include a concern that covering content and conveying what is "important" means that time available for thinking (personal construction) is pushed down, and often off, the agenda; the "efficiency" of presenting content in a transmissive rather than an interpretive mode (teaching is thought to be more important than learning!); the establishing of different "rules" and values in both classroom and teacher education contexts; and establishing conditions of trust to allow risks to be taken and respect for the views and experiences of all participants to develop.

CONCLUSION

Difficulties should not be underestimated for those who wish to apply constructivist ideas to teaching/learning situations. The case examples in this chapter are encouraging signs that many teachers and teacher educators are willing to face these issues. Also, there has been an encouraging shift in the way research is being conducted within this constructivist area. There is an increasing respect for teacher knowledge and research in areas of teaching and learning, metacognition, and student intuitions. There is a recognition that progress requires collaboration and action research and that many teachers are reviewing their professional role. Perhaps we are beginning to change the conditions of teaching and our expectations of the teaching role. Constructivist ideas could provide the rationale and guidelines for such a reform of teaching.

REFERENCES

Baird, J. R., & Mitchell, I. J. (1986). *Improving the quality of teaching and learning: An Australian case study—The PEEL project*. Melbourne, Australia: Monash University.

Baird, J. R., & Northfield, J. R. (1990). *PEEL at Laverton Secondary College 1990: A report based on interviews held with nine PEEL teachers*. Unpublished paper, Monash University, Melbourne, Australia.

Baird, J. R., & Northfield, J. R. (1992). *Learning from the PEEL experience.* Melbourne, Australia: Monash University.

Department of Employment, Education and Training. (1989). *Discipline review of teacher education in mathematics and science* (3 vols.). Canberra, Australia: Australian Government Publishing Services.

Erickson, G. L. (1988, June). *Processes and products from the (SI)² project: Anatomy of a collaborative approach.* Paper presented at the meeting of the Canadian Society for the Study of Education, Windsor, Ontario, Canada.

Erickson, G. L. (1991). Collaborative inquiry and the professional development of science teachers. *Journal of Educational Thought, 25,* 228–245.

Erickson, G. L., & MacKinnon, A. (1991). Seeing classrooms in new ways: On becoming a science teacher. In D. Schön (Ed.), *The reflective turn: Case studies of reflection in and on practice* (pp. 15–36). New York: Teachers College Press.

Fullan, M. (1988). *Change processes in secondary schools: Toward a more fundamental agenda.* Unpublished paper, Teacher Context Center, Stanford University, Palo Alto, CA.

Gunstone, R. F., & Northfield, J. R. (1988, April). *In-service education: Some constructivist perspectives and examples.* Paper presented at the annual meeting of the American Educational Research Association, New Orleans, LA.

Gunstone, R. F., Slattery, M., & Baird, J. R. (1989, April). *Learning about learning to teach: A case study of pre-service teacher education.* Paper presented at the annual meeting of the American Educational Research Association, San Francisco, CA. (ERIC ED 307150)

Hewson, P. W. (1981). A conceptual change approach to learning science. *European Journal of Science Education, 3,* 383–396.

Northfield, J. R., & Gunstone, R. F. (1983). Research on alternative frameworks: Implications for teacher education. *Research in Science Education, 13,* 185–191.

Schön, D. (1983). *The reflective practitioner: How professionals think in action.* New York: Basic Books.

Schön, D. (1987). *Educating the reflective practitioner.* San Francisco, CA: Jossey-Bass.

Schön, D. (1988). Coaching reflective teaching. In P. Grimmett & G. Erickson (Eds.), *Reflection in teacher education* (pp. 19–29). New York: Teachers College Press.

Walker, R. (1989). Monash University Diploma in Education "Stream Three" (Science)—A case study. In Department of Employment, Education and Training, *Discipline review of teacher education in mathematics and science* (Vol. 3, pp. 42–52). Canberra, Australia: Australian Government Publishing Services.

CHAPTER 18

Diagnosis of Teachers' Knowledge Bases and Teaching Roles When Implementing Constructivist Teaching/Learning Approaches

Brian Hand
LaTrobe University, Bendigo, Australia

Constructivist approaches to teaching and learning science have gained acceptance over the last decade. Concomitantly, there is a need for science teachers to change and adopt new pedagogical knowledge (Millar & Driver, 1987) and to examine their current practices in order to implement student-centered rather than teacher-centered approaches. In-service education programs are one means whereby teachers can be provided with opportunities to change their existing pedagogical knowledge to accommodate constructivist teaching/learning approaches. As teachers examine the processes involved in constructivist teaching/learning approaches they may undergo their own conceptual change and may engage in conceptual change teaching (Anderson & Smith, 1987).

Changes from previously implemented practices to those required for constructivist teaching/learning approaches do not take place all at once, and teachers experience a number of stages in implementing and adopting new approaches. For example, Simon and Schifter (1987) describe three stages of knowledge or approaches through which teachers go when changing to constructivist approaches: first, no knowledge of or use of constructivist epistemology; second, a mechanical application of constructivist approaches; and third, a focus on student learning rather than teaching behaviors. Based on the work of Habermas (1972), Grundy (1987) identified three different interests through which teachers implement new approaches. Initially, teachers have technical interests and follow procedures outlined to them. As teachers gain confidence, they have practical interests and develop greater insights into the consequences of implementing new approaches. The final interest is emancipatory when the goal is to empower individuals or groups to be more autonomous.

This chapter comprises three further sections. In the first section attention is given to the different types of knowledge bases and roles that teachers must develop and take on if they are to effectively implement constructivist teaching/learning approaches. The second section describes how teachers' knowledge bases and roles can be observed and diagnosed as constructivist teaching/learning approaches are implemented. The third section of this chapter describes an in-service model comprising five stages showing how a teacher's knowledge bases and teaching roles develop at each stage of the in-service program for implementing constructivist teaching/learning approaches.

TEACHERS' KNOWLEDGE BASES AND ROLES

Teachers' knowledge bases and roles in the classroom can be examined to diagnose changes taking place as constructivist approaches are successfully implemented.

In order for teachers to successfully implement constructivist approaches, they need to develop different *knowledge bases.* Constructivist approaches to learning require the teacher to have a much broader view of subject-matter knowledge, with a changing emphasis from the transmission of decontextualized content knowledge to allowing individual construction of contextualized knowledge (Prawat, 1993). According to Wilson, Shulman, and Richert (1987), subject-matter knowledge is defined not only as the substantive structures for content knowledge, but also as syntactic structures of the discipline that "involve knowledge of the ways in which the discipline creates and evaluates new knowledge" (p. 118). By focusing on linking concepts and their relationships in a specific context, teachers will be able to promote "[conceptual] frameworks for structuring and interpreting experiences" (Carter & Doyle, 1987, p. 149). As constructivist approaches place importance on the process of assimilating or accommodating new concepts into an individual's existing conceptual framework, there is a need to consider the viability of the meaning of subject-matter knowledge put forward by Wilson and colleagues (1987). Research in science classrooms (for example, Gallagher, 1991; Tobin & Fraser, 1989; Tobin & Gallagher, 1987) has suggested that teachers focus on the rote learning of facts and algorithms whereby students gain decontextualized content knowledge. This type of content knowledge is one part of subject-matter knowledge described by Wilson and colleagues (1987) and lacks the focus on concepts and syntatic structures that they believe are important.

In order for teachers to encompass the notion of concepts and syntatic structures in their teaching of content, they must have a greater understanding and awareness of the conceptual knowledge associated with the topic being taught (Bennett, 1988), and a clear conception of what ideas or concepts are central to the discipline and how they are related to one another (Prawat, 1989). This conceptual knowledge is derived from the relationship between content knowledge and the context within which the knowledge is constructed and recognized, and has the potential for illuminating aspects of the physical and social world that otherwise would go unnoticed or unappreciated.

In reporting a series of workshops to address the lack of focus on teaching conceptual knowledge, Bowden (1988, p. 260) stated that the teachers struggled when asked to go beyond a description of content areas (i.e., they struggled with the concepts and their links). The teachers realized that, when focussing on teaching concept knowledge, their normal teaching practices were incompatible with the desired outcomes. To bring the two more closely together required change on the "teacher's part towards a view of teaching as changing conceptions" (Marton & Ramsden, 1988, p. 276), the adoption of new pedagogy (Millar & Driver, 1987), and the development of pedagogical knowledge centered on the teaching of conceptual knowledge (i.e., pedagogical concept knowledge). Such pedagogical concept knowledge involves the use of negotiation and group work, interpretative discussions, and wait-time.

In adopting constructivist approaches, teachers have to take on different roles in the classroom, which may be characterized by Grundy's technical, practical, and emancipatory interests. Adopting a technical interest, teachers develop a "teacher-as-technician" role in which they implement the new approach in a mechanical, rule-following manner. Adopting practical interests, teachers play the role of "teacher-as-facilitator," with greater emphasis on exploring ways and means of promoting active student participation within the learning process. Involved in this process is the development of each teacher's own conceptual understanding of the changing pedagogy. Adopting emancipatory interests, teachers develop the role of "teacher-as-empowerer," and students are encouraged to become problem-setters rather then problem-solvers (i.e., to ask and solve questions or problems, rather than just to respond to problem situations).

OBSERVING TEACHERS' KNOWLEDGE BASES AND TEACHING ROLES

Classroom observations, teachers' journal descriptions of metaphors (that is, descriptions of implicit comparisons between teaching and other activities), and semi-structured interviews can be used for documenting and diagnosing changes made by teachers as they implement constructivist approaches.

When a teacher uses constructivist approaches, learning is negotiated between teacher and students, and the number of activities controlled by the teacher decreases. As a teacher's classroom role becomes more facilitative, there should be an observable increase in the joint negotiation of ideas with students and opportunities for students to pose questions. Carefully structured classroom observations can be used as a means to determine the degree to which the students become active, sharing partners in the learning process.

In relation to constructivist teaching/learning approaches, Marshall (1990) recommends that teachers move away from a workplace metaphor that includes an authority figure who has status and power. For Tobin (1990), metaphors are used to conceptualize teaching roles. Metaphors can change in the process of the teacher changing his or her role in the classroom and, once a role has changed, the beliefs about that role can be deemed to apply no longer. Consequently, when teachers change their pedagogical content knowledge to include pedagogical concept knowledge there will be changes in their beliefs, which are detected in the metaphors used to describe their teaching. As part of the process of keeping a journal, teachers should be asked to record one or more metaphors to describe their teaching prior to the in-service program, on completion of the teaching, and several months later. Interviews also can be used as an effective means to examine teachers' knowledge bases and roles provided that they are conducted prior to and on completion of the in-service program.

A MODEL FOR DIAGNOSING TEACHERS' KNOWLEDGE BASES AND TEACHING ROLES

The proposed model for diagnosing teacher change was developed from an extended in-service education program with eight secondary school science teachers. As a consequence of the Victorian Ministry of Education's proposed curriculum changes, these teachers were expected to become more involved than previously with teaching/learning strategies that focused much more on the learner. In order to become familiar with constructivist approaches, the teachers volunteered to participate in an 18-month in-service program based on Driver and Oldham's (1986) curriculum model for constructivist teaching/learning approaches. (See also Driver and Scott, Chapter 8, this volume.) Over the period of four months before the in-service activities began, each teacher was observed in the classroom and interviewed in relation to his or her philosophy of teaching. The second phase of four months consisted of a series of meetings for discussing current teaching practices, reading and reviewing literature relating to constructivist approaches (in particular, new pedagogical approaches required), trying out new skills, and planning units for implementation. The third phase, spread out over 10 months, involved each teacher's implementing the new approaches within the classroom. During the in-service program, regular planning, observation, and debriefing sessions were held with each teacher.

Based on my involvement with the teachers and the data collected, a model was developed for diagnosing changes in teachers' knowledge bases and roles when they participated in an in-service program aimed at encouraging and fostering the implementation of constructivist teaching/learning approaches in the classroom. The model presented in Figure 18.1 consists of five stages, each of which can be identified by examining changes of the teachers' knowledge bases and roles. The first four stages of the model closely match the four phases described by Driver and Oldham (1986) for a constructivist approach to curriculum development.

Stage 1
Identification of Teacher's Knowledge of Classroom Practice

Initially, the teachers were asked to describe successful strategies that they implemented to promote student learning and, hence, to define their own pedagogical content knowledge as a beginning point for change. Prior to the in-service course, classroom observations indicated that all teachers taught in an information-transfer mode, which had an emphasis on ensuring that the correct scientific content knowledge was given to the students. The major concern expressed by teachers was the need to maintain control of the students and the content knowledge to be covered. These concerns were reflected in the managerial roles adopted by the teachers; for example, metaphors used by teachers included "ring master" and "lecturer." Classroom observations indicated that two-thirds of the lessons were teacher-controlled and that student-controlled activities were of low order and unlikely to promote higher-order thinking (Hand & Treagust, 1994).

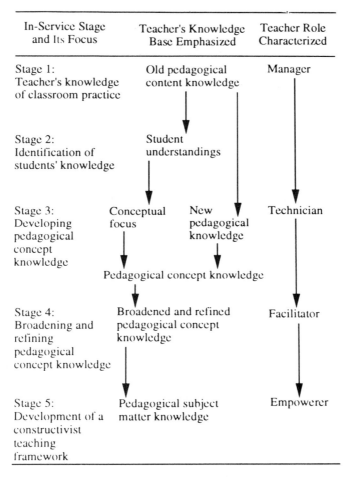

Figure 18.1. Five-Stage In-Service Model for Diagnosing the Implementation of Constructivist Teaching/Learning Approaches

Stage 2
Teacher Identification of Students' Knowledge of Science

All the teachers were asked to try out an initial strategy aimed at determining students' conceptions of a particular topic of their choosing and, where necessary, assistance was given in using appropriate strategies, for example, a free-write process. Teachers were not asked to change their pedagogical knowledge to achieve this task, but only to conduct one lesson to ascertain the students' ideas. As a result of having found out that students had different conceptions from what teachers believed that students should have, the teachers realized that their pedagogical knowledge was inadequate to address the problem of challenging the students' ideas; teachers had observed a mismatch between the students' and their own ideas about the topic, but were unsure how to deal with this situation.

Stage 3
Developing Pedagogical Concept Knowledge

Having recognized the need to change their classroom practices, the teachers were given a set of readings about constructivist teaching/learning approaches. These readings formed the basis of discussions for determining which new pedagogic skills were needed when implementing new approaches; the skills included improving questioning, conducting interpretative discussions, and using small-group and large-group work. In conjunction with this focus on developing new skills, particular emphasis was placed on the need to focus on conceptual knowledge rather than on content knowledge, a process that was both new and difficult for the teachers.

Due to unfamiliarity with the new skills, the teachers were encouraged to follow those implementation processes discussed during the in-service education sessions. Initially, the teachers acted as technicians (Hand & Treagust, 1991) in that they did not attempt to explore the effects of the new approaches, but instead concentrated on ensuring that they followed the implementation processes. According to classroom observations, the two teachers who remained at stage 3 at the completion of the in-service program only had a 20 percent decrease in the number of activities they controlled. Although these teachers were adopting student-centered approaches, they were reluctant to give much more control to the students, and used the metaphors of a "director of discussion" and "pilot of an aircraft" to describe their teaching.

Stage 4
Broadening and Refining Pedagogical Concept Knowledge

The intention at this stage was to allow the teachers to assimilate and/or accommodate their own knowledge of pedagogy and science addressed in stage 3 in order to broaden and refine their existing pedagogical concept knowledge. Following the tryout sessions conducted in stage

3, when new pedagogical skills (such as defining student concepts) were practiced, constructivist teaching/learning approaches were discussed, and each teacher was asked to choose a topic with which to use the new approach. Teachers were given assistance with planning and implementing the topics chosen. Much of this time was used to identify concepts and examine possible learning pathways students could take. As part of the in-service program, the teachers were asked to record the planning sessions and observations of implementing the approach, particularly the students' concepts, the processes used in developing those concepts, and the reactions of the students to the new approach. The journals that the teachers kept and the debriefing sessions with the author encouraged the teachers to reflect on their experiences.

As the teachers constructed pedagogical concept knowledge, there was an associated change in role from technician to facilitator (Hand & Treagust, 1991). The teachers' confidence in using the constructivist strategies increased as students became more positive toward science when opportunities were given to become more involved in their own learning. Teachers experimented with a range of ways of applying the new pedagogic skills, such as using newspaper articles, story writing, and using scenarios to begin a topic, rather than using the free-write process. Classroom observations indicated a reduced overall emphasis on teacher-controlled activities (34 percent), with a decrease of 32 percent from when classroom observations were made prior to the in-service education program (Hand & Treagust, 1994). This change was reflected in the metaphors "tour guide" and "house renovator" that the teachers used to describe their practice.

Stage 5
Development of a Constructivist Teaching Framework

In the final stage of the in-service model, the focus of teaching was no longer conceptual knowledge, but rather subject-matter knowledge, which, as described previously, consists of content and concepts. Teachers' pedagogical knowledge changes to allow students to become problem-setters and not problem-solvers and to provide opportunities for students to participate in setting the direction for the topic, by students' developing and using research methodologies and by posing appropriate questions and strategies to rationalize solutions. At this stage, the role adopted by teachers is that of an empowerer rather than a facilitator who provides problems for students to investigate.

CONCLUSION

This chapter has described a five-stage model of in-service education for implementing new teaching approaches, particularly those involving constructivism. Science teachers' practices can be observed, documented, and diagnosed by analyzing teachers' knowledge bases and teaching roles within the classroom.

The model presented in this chapter has been developed as the result of a single study and is an initial attempt to diagnose teachers' knowledge and roles when adopting constructivist teaching/learning approaches. The eight teachers who participated in and completed the in-service program differed in the stage they reached. Two teachers were at the third stage, one was in transition between stages 3 and 4, and five reached stage 4. More research must be conducted into ways to monitor the development of teachers through the various stages of the model. For example, a questionnaire addressing both teachers' and students' perceived and preferred views of constructivist learning environments (Taylor, Fraser, & White, 1994) could be adapted to allow teachers to self-monitor their progress toward adopting constructivist approaches. Emphasis must be placed on developing instruments that allow in-service personnel, researchers, and teachers the opportunity to monitor progress easily and effectively.

REFERENCES

Anderson, C. W., & Smith, E. L. (1987). Teaching science. In V. Richardson-Koehler (Ed.), *Educator handbook: A research perspective* (pp. 84–111). London, England: Longman.

Bennett, N. (1988). The effective primary school teacher: The search for a theory of pedagogy. *Teaching and Teacher Education, 4*, 19–30.

Bowden, J. (1988). Achieving change in teaching practices. In P. Ramsden (Ed.), *Improving learning: New perspectives* (pp. 255–267). London, England: Kogan Page.

Carter, K., & Doyle, W. (1987). Teachers' knowledge structures and comprehension processes. In J. Calderhead (Ed.), *Exploring teachers' thinking* (pp. 147–160). London, England: Cassell Educational Limited.

Driver, R., & Oldham, V. (1986). A constructivist approach to curriculum development in science. *Studies in Science Education, 13*, 107–112.

Gallagher, J. J. (1991). Prospective and practicing secondary school science teachers' knowledge and beliefs about the philosophy of science. *Science Education, 75*, 121–133.

Grundy, S. (1987). *Curriculum: Product or praxis.* London, England: Falmer Press.

Habermas, J. (1972). *Knowledge and human interests.* London, England: Heinemann.

Hand, B., & Treagust, D. (1991, April). *From teacher-as-technician to teacher-as-facilitator: A study of a professional development initiative involving teaching for conceptual change.* Paper presented at the annual meeting of the American Educational Research Association, Chicago, IL.

Hand, B., & Treagust, D. (1994). Teachers' thoughts about changing to constructivist teaching/learning approaches within junior secondary science classrooms. *Journal of Education for Teaching, 20,* 97–112.

Marshall, H. H. (1990). Beyond the workplace metaphor: The classroom as a learning setting. *Theory into Practice, 29,* 94–101.

Marton, F., & Ramsden, P. (1988). What does it take to improve learning? In P. Ramsden (Ed.), *Improving learning: New perspectives* (pp. 268–286). London, England: Kogan Page.

Millar, R., & Driver, R. (1987). Beyond processes. *Studies in Science Education, 14,* 33–62.

Prawat, R. S. (1989). Teaching for understanding: Three key attributes. *Teaching and Teacher Education, 5,* 315–328.

Prawat, R. S. (1993). The value of ideas: Problem versus possibilities in learning. *Educational Researcher, 22,* 5–16.

Simon, M. A., & Schifter, D. (1987). *Teacher education from a constructivist perspective: The educational leaders in mathematics project.* Washington, DC: National Science Foundation.

Taylor, P., Fraser, B., & White, L. (1994, April). *CLES: An instrument for monitoring the development of constructivist learning environments.* Paper presented at the annual meeting of the National Association for Research in Science Teaching, New Orleans, LA.

Tobin, K. (1990). Metaphor in the study of teachers' professional knowledge. *Theory into Practice, 29,* 122–127.

Tobin, K., & Fraser, B. (1989). Barriers to higher level cognitive learning in high school science. *Science Education, 73,* 659–682.

Tobin, K., & Gallagher, J. J. (1987). What happens in high school science classrooms? *Journal of Curriculum Studies, 19,* 549–560.

Wilson, S. M., Shulman, L. S., & Richert, A. E. (1987). "150 different ways" of knowing: Representations of knowledge in teaching. In J. Calderhead (Ed.), *Exploring teachers' thinking* (pp. 104–124). London, England: Cassell Educational Limited.

CHAPTER 19

Implementing Teacher Change at the School Level

James J. Gallagher
Michigan State University, East Lansing, United States

In this chapter, I describe the Support Teacher Program which was designed to improve science and mathematics teaching at the junior high school level. The chapter consists of three parts: a discussion of how widely held cultural beliefs about science teaching give rise to difficulties in engaging students in learning science; the design of the Support Teacher Program to overcome these difficulties; and resulting changes in the Support Teachers' beliefs and teaching, in the climate of the departments and the school, in their peers' beliefs about teaching, and in students' learning.

SCIENCE TEACHING AND TEACHERS' BELIEFS

For many science teachers, the central task in teaching secondary school science has been perceived as presenting science content prescribed in a syllabus or completing the text that has been approved by school officials or an external agency. "Covering the content" is the common phrase that is used by science teachers to refer to this belief about their work. Teaching often is viewed as "getting the work done" and the work is to "cover the content" defined in the syllabus or text (Tobin & Gallagher, 1987). This viewpoint is part of the belief system of many who teach science at both secondary and tertiary levels. It is reinforced in conversations that occur among teachers and professors of science, and by many systems of testing and "accountability." The belief system appears to be commonplace and can be considered to be a central part of the "culture of science teaching" that prevails in schools and universities around the world. There are exceptions. Exemplary teachers place less emphasis on "coverage," showing greater concern for students' understanding of scientific content, their capability to apply it to real problems that exist beyond the walls of the school, while nurturing scientific habits of mind and positive attitudes toward science and self (Tobin & Fraser, 1987, p. 209).

Many of the difficulties that we face today in educating young people in science are consequences of the actions of teachers and university professors whose work is guided by the narrow beliefs that limit the vision of their work to presenting content for students to learn. Further, these educators do not assume responsibility for assisting students in understanding the subject matter, in learning how to apply it, in developing positive attitudes about science and self, or in developing the habits of mind that characterize science. This narrow vision of science teaching fails on a most essential point by not acknowledging the depth and complexity of the work of science teachers, without which effective teaching is unlikely to occur.

A number of difficulties derive from this narrow viewpoint about science teaching:

- When teachers do not acknowledge the complexity of their work, they are unlikely to give attention to teaching and learning tasks that result in understanding, application, and development of attitudes and habits of mind that are part of the intended outcomes of science teaching.
- When attention is not given to helping students achieve understanding of the subject matter of science, students typically memorize elements of subject matter in an unconnected way and quickly forget what they have learned.
- When attention is not given to developing understanding, students are unable to appreciate the logical order of scientific knowledge, and are more likely to perceive scientific knowledge as illogical and unconnected.
- When attention is not given to applications of scientific knowledge that have some current utility to students, science is seen as abstract, irrelevant, and of little value.
- When attention is not given to nurturing positive attitudes, students can develop a distaste for science and a decreasing interest in and appreciation for science each year of study.
- When attention is not given to nurturing scientific habits of mind, students are unlikely to address problems systematically, continuing to jump to conclusions when evidence is sketchy, instead of withholding

judgment while seeking more information. Further, they will be unlikely to develop the "healthy skepticism" that is needed to be analytical about claims that are made in the media, advertising, political rhetoric, and other forms of propaganda.

However, these beliefs about the teacher's work as covering content are deeply ingrained and are reinforced internally by the social system within many science departments and externally by evaluation systems and syllabi that emphasize factual knowledge from a broad scope of content. Therefore, they are difficult to change, thus contributing to the failure of many attempts at improving science teaching.

CHANGING THE CULTURE IN MIDDLE-SCHOOL SCIENCE AND MATHEMATICS DEPARTMENTS

In February 1988, the Institute for Research on Teaching at Michigan State University began a program to improve teaching and learning in middle-school science and mathematics in conjunction with the American Federation of Teachers and the Toledo (Ohio) Public Schools. An important part of this program was to change the culture within middle schools, at the same time as teachers developed a new vision of teaching and new skills to implement that vision. The intent was to help teachers to think differently about their work, and to develop new skills to make it possible to teach more effectively.

Details of the Support Teacher Program are described in Gallagher, Lanier, and Kerchner (1993). In brief, the program was designed to enable the Support Teachers, who were among the district's best teachers, to

- Improve their own teaching and their students' learning; help their colleagues (departmental peers), in a supportive, nonevaluative context, improve their teaching and their students' learning through both individual and departmental effort; and continue to study to keep abreast of, and adapt, new developments that can improve their effectiveness in the other two dimensions of their role.
- Work in pairs in each school—one in science and one in mathematics—teaching half-time and working in the support role half-time.
- Develop new understandings and skills informed by recent research about teaching, learning, and assisting peers in improving their teaching and learning. Thus, learning about recent research and translating it into practice would be a centerpiece of the cooperative relationship among the research university, the school district, and the teachers' union.

- Improve their own teaching in order to provide effective models of teaching for their peers and to learn firsthand about the difficulties and traumas that can accompany attempts at changing teaching. To help Support Teachers experience the change process in a personal way, we began our program with a strong emphasis on analysis and improvement of their teaching and their students' learning.
- Learn skills in working with peers, including how to observe teaching and give feedback, how to organize and conduct staff meetings, and how to assist peers in changing their teaching.

PROGRAM DEVELOPMENT FOR SUPPORT TEACHERS AND THE TEACHING MODEL

The plan begun in February 1988 was designed to prepare the eight Support Teachers to assume a new role in their respective departments in helping their peers to improve their teaching and students' learning in science and mathematics. The eight Support Teachers had 15–23 years of experience, were identified by their administrators and peers as "good" teachers, and were confident in their ability as teachers, believing that their students were learning effectively. However, all were concerned that they were not motivating their students adequately, especially minority students, who comprised 40 percent of the school district's population. Initially, training sessions of two days' duration were held approximately every three weeks from February to May 1988, followed by a week-long program in August, just prior to the Support Teachers' initiation into their new role. The purposes of these sessions were to help Support Teachers improve their own teaching and to develop skills in working with their peers to nurture their improvement. To assist the Support Teachers, we provided a notebook of readings on recent research on teaching and learning in science and mathematics and supervision of teachers to support professional growth. Through seminars, Support Teachers were helped to make connections between the research and their own work. We also observed them teaching their own classes to model the use of observation and feedback in improving teaching and learning.

Through readings and seminars, Support Teachers became more aware of how their own students' misunderstanding of scientific concepts would cause teachers to question their own teaching and why students were not learning the subject matter as anticipated. Teachers asked, "Why are my students not learning what I thought they were?" and "What can I do differently to help them understand?" From these questions, we developed a model to guide teaching and learning derived from

Driver's (1989) work (Gallagher, 1993). Concurrently, the Support Teachers studied how to work with their peers. They read and analyzed research reports and engaged in practical experiences including receiving and providing supervision based on detailed classroom observations.

SOME OUTCOMES OF THE SUPPORT TEACHER PROGRAM

Outcomes of the Support Teacher Program are described in terms of changes that have taken place: changes in Support Teachers' beliefs and teaching; changes in departmental and school climate; changes in peers' beliefs and teaching; and changes in student learning.

Changes in Support Teachers' Beliefs and Teaching

Changes in the beliefs of the Support Teachers were monitored through successive administrations of the Science Teaching Style Inventory (Madsen & Gallagher, 1992), formal interviews, and qualitative observational data. The Teaching Style Inventory and interviews were administered four times during the first two and one-half years of the program and yearly thereafter. The data were used to cross-validate our understanding of teachers' beliefs by categorizing them as being traditional (didactic), transitional, or nontraditional (constructivist). This categorization was based on teachers' statements and responses to specific questions about lesson content, learning tasks, and social organization for learning (Madsen & Gallagher, 1992). All eight Support Teachers showed substantial changes in their beliefs from traditional/didactic beliefs when the project began in 1988 through to the transitional category, and then entering the nontraditional, constructivist category after a period of two years (Madsen & Gallagher, 1992). Classroom observations confirmed this same progression.

The change in beliefs of the Support Teachers appears to be essential to the changes in classroom content and interactions with their students, and to their work with their peers. Moreover, the change in beliefs occurred slowly, over several months, in spite of ongoing intervention at a moderate level of intensity (i.e., approximately two days every three weeks). Further, it took longer for the skills and techniques necessary to implement the new vision to be sharpened so that the Support Teachers were confident of their capability to use them effectively in class. During the past four years, the shift in both beliefs and actions advanced their constructivist teaching toward a new category called constructivist-inquiry teaching (Gallagher & Parker, 1994).

Changes in Departmental and School Climate

There has been a significant change in the intellectual climate in all science and mathematics departments, as a consequence of the Support Teacher Program. Prior to and during the first four months of the program, there was little conversation about teaching and learning in these eight departments. Departmental meetings were rare, and when they did occur, logistical issues such as schedules, supplies, and materials dominated. Teachers did not talk about students' understanding of subject matter, or how to improve it. Informal conversations tended to be about at-risk students who were perceived as troublemakers, or about school funding and other problems in the larger society, such as parental divorce and drugs, which were affecting students. Much of the emphasis was negative. Little conversation occurred about actions that teachers could take that would affect perceived problems within the school, the department, or the teachers' own classrooms.

Beginning in September 1988, the Support Teachers organized regular departmental meetings with agendas that addressed specific actions to change teaching and improve learning. The "new vision" of teaching and learning, and the techniques to achieve it, became a more common part of the discourse in departments both in scheduled meetings and informal conversations between Support Teachers and their peers (Bettencourt & Gallagher, 1989, 1990). In addition, Support Teachers began to work individually with teachers to assist them in improving particular aspects of their teaching.

The efficacy of this program is illustrated by comments from a few teachers who taught in these four junior high schools with the Support Teacher Program, and who have been reassigned to other schools that do not have it. They lament the loss of interaction with peers that they previously enjoyed. Further, many teachers in departments other than science and mathematics in the four junior high schools with the Support Teacher Program expressed their desire for a similar program in their departments because it gives a better sense of community among teachers, and it provides assistance in addressing difficult problems of teaching and learning.

Association of this project with both the district and the union was important in the success of the project. The district made it possible for the Support Teachers to have a reduced teaching load (three classes per day instead of the normal load of six classes), which provided time to carry out their new responsibilities. In addition, each of the four buildings where the Support Teacher program is in place has established a science and mathematics resource room as a work space for science and mathematics teachers, and as a place to house teaching materials and journals. These costs are borne by the district. The support of the teachers' union also has been a critical factor in the program's continuance and success. The union president was a co-initiator of the program and he demanded its continuation during a serious fiscal crisis in the district. Moreover, the union leaders have been very supportive of the program in the few conflicts that arose between Support Teachers and some of their most resistant peers.

Changes in Peers' Beliefs and Teaching

Data from teaching peers' classrooms, interviews, and successive completions of the Learning Style Inventory show that beliefs and teaching activities have changed markedly as the program has continued. The peer teachers also view teaching differently than in the past and are using more effective teaching techniques in their classes. However, it must be emphasized that some teachers in each department have changed slowly.

Both Support Teachers and their peers have reduced the amount of content coverage. As people begin to assist students with understanding and applying scientific and mathematical knowledge, they immediately are confronted with the realization that they cannot "cover" as many units during the year. When the program began, we negotiated the issue of coverage with the district curriculum director, reducing the number of science units from eight to four per year. At first the teachers felt guilty for eliminating half of the traditional content from junior high school science. However, as they began to strive for understanding and application of knowledge, they soon realized that covering eight units per year had many detrimental effects on students' learning and self-esteem, and their perceptions of scientific logic. Parallel changes occurred in mathematics.

Two other attributes readily evident in the classrooms of both the Support Teachers and their peers are the social organization of instruction and the learning tasks that teachers provide for students. There now is a much greater balance between whole-class instruction, cooperative group work, and individual seatwork compared with the way in which seatwork and whole-class instruction dominated in the past. Moreover, students do more of the talking, and teachers talk less, in these classes than before the program began to affect practice. Further, there is much more student writing of paragraphs and essays as part of science and mathematics classes. Explanations and interpretations of data from observations are much more evident in science classes. There are fewer examples of students' responding with one-word answers, because the questions teachers ask of students cannot be answered in this way, either during oral discourse or when written work is required. There are many more instances in classes where students are working in groups to answer questions that require group discussion, analysis, group writing, and public reporting. Also, there is much greater emphasis now on using modes of expression other than words, including pictures, graphs, and models, to represent ideas and relationships in science. Constructivist approaches to teaching are now common, whereas they rarely occurred in these schools prior to this program.

Changes in Student Learning

The limited data available provide compelling evidence of the effectiveness of the program in increasing student learning, improving student attitudes toward science, and nurturing positive habits of mind. Because the district does not use a standardized test in science at the middle-school level, local science achievement tests that are relevant to the topics taught by the Support Teachers were created and administered. Performance of students in the classes of two teachers in the same building were compared using these tests. For example, large differences in achievement existed between the class of Mr. X, who had adopted and applied the principles of the Support Teacher Program, compared with that of Mr. R, who had been resistant to change. On a test with a maximum score of 60 points, Mr. X's students scored an average of 42 points while Mr. R's students scored 23 points (Madsen & Gallagher, 1992). While the difference in scores is impressive, an even more profound difference was found in the habits of mind of the students in the two classes. On two questions requiring students to solve novel problems, all of Mr. X's students attempted the problems, while only 29 percent of Mr. R's students attempted one of the problems and 66 percent attempted the second problem. Such differences have important implications for future success in learning and problem solving.

Another data source pertaining to students' habits of mind is the teachers in high schools that receive students from different junior high schools. In Toledo, there are eight junior high schools and, until recently, only four were participating in the Support Teacher Program. The city's four high schools receive a mixture of students from these junior high schools. On several occasions, Support Teachers have been in meetings with science teachers from these high schools and have received very positive comments about their former students.

One additional compelling piece of evidence comes from the standardized achievement tests given in mathematics. In 1990, eighth-grade students whose teachers had been in the program for one year prior to that school year (1988–1989) achieved at a level below national norms for the test (49 percent below the thirty-sixth percentile, 5 percent above the eighty-fifth percentile). A year later, when the teachers had two years experience in the Support Teacher Program, students' achievement was at the national norms (35 percent below the thirty-sixth percentile, 15 percent above the eighty-fifth percentile). This represents an important gain in just one year. In following years, the improvements have continued to grow even though the mathematics teachers, like the science teachers, have taught fewer topics, while emphasizing understanding and application. These data give some additional justification to the principle that "less is better" (Rutherford & Ahlgren, 1989).

CONCLUSION

Skeptics might be unconvinced about the answer to the question, "Does this Support Teacher Program affect students' learning and attitudes?" However, the available data all point to beneficial effects. Support Teachers and their peers have changed their beliefs about teaching and learning and are employing different methods in their classes. The climate of their departments has changed markedly, and the discourse among teachers focuses on how to improve teaching and learning. The work of students includes a greater emphasis on activities that produce understanding of science and the capability and propensity to apply it in solving problems and seeking further understanding. Finally, the students appear to be attaining a deeper understanding of scientific principles, as well as desired habits of mind. Also attesting to the success of the program is the recent expansion of the program to two additional junior high schools in Toledo, with the consent of the teachers in the two schools, the central administration of the district, and the union.

ACKNOWLEDGMENTS

The key participants in initiating this program were the president of the Toledo Federation of Teachers, who also is a vice-president of the American Federation of Teachers, an associate superintendent of Toledo Public Schools, eight teachers from four junior high schools in Toledo who were to take on a new role identified as Support Teachers, and five staff members of the Institute for Research on Teaching, including the author. I am deeply indebted to these colleagues for their outstanding work on this project.

REFERENCES

Bettencourt, A., & Gallagher, J. J. (1989, March-April). *Helping science teachers help science teachers: A study of change in junior high school science.* Paper presented at the annual meeting of the National Association for Research in Science Teaching, San Francisco, CA.

Bettencourt, A., & Gallagher, J. J. (1990, April). *Changing the conversation: When science teachers start talking about instruction.* Paper presented at the annual meeting of the National Association for Research on Science Teaching, Atlanta, GA.

Driver, R. (1989). Changing conceptions. In P. Adey (Ed.), *Adolescent development and school science* (pp. 79–103). London, England: Falmer Press.

Gallagher, J. J. (1993). Secondary science teachers and constructivist practice. In K. Tobin (Ed.), *The practice of constructivism in science education* (pp. 181–191). Washington, DC: American Association for the Advancement of Science.

Gallagher, J., Lanier, P., & Kerchner, C. (1993). Toledo and Poway: Practicing peer review. In C. Kerchner & J. Koppich (Eds.), *A union of professionals: Labor relations and educational reform* (pp. 158–176). New York: Teachers College Press.

Gallagher, J., & Parker, J. (1994). *A category system for analysis of science teaching.* Unpublished working paper, Project Salish, Michigan State University.

Madsen, A., & Gallagher, J. J. (1992). Improving learning and instruction in science classes through support teaching. Unpublished working paper, Michigan State University.

Rutherford, J., & Ahlgren, A. (1989). *Science for all Americans.* Washington, DC: American Association for the Advancement of Science.

Tobin, K., & Fraser, B. (Eds.). (1987). *Exemplary practice in science and mathematics education.* Perth, Australia: Key Centre for School Science and Mathematics, Curtin University of Technology.

Tobin, K., & Gallagher, J. J. (1987). Teacher management and student engagement in high school science. *Science Education, 71,* 535–556.

About the Editors and the Contributors

THE EDITORS

Reinders Duit is Professor of Physics Education and a member of the Physics Education Group at the Institute for Science Education (IPN) at the University of Kiel, Germany. He earned his Ph.D. in 1972 investigating long-term changes of students' knowledge structures in the domain of heat phenomena. His main research interest has been difficulties in learning basic science concepts and incorporating constructivist ideas into mainstream secondary school classrooms. His research includes studies on students' learning processes in the domains of electricity, energy, heat, and, more recently, chaotic systems.

Barry J. Fraser is Professor of Education and Director of the national Key Centre for School Science and Mathematics at Curtin University in Perth, Australia. Prior to coming to Curtin, he taught at Macquarie and Monash universities, earning his Ph.D. from Monash in 1976, and was a high school science and mathematics teacher in Australia. His professional interests are in classroom learning environments and in-service teacher education. He is currently President of the National Association for Research in Science Teaching in the United States.

David F. Treagust is Associate Professor of Science Education at the Science and Mathematics Education Centre at Curtin University in Perth, Australia. Prior to joining the Curtin faculty as a lecturer in 1980, he held a postdoctoral appointment at Michigan State University, completed his Ph.D. at the University of Iowa in 1978, and taught high school science in Australia and England. His research interests are related to understanding students' ideas and how those ideas can be implemented as part of the regular curriculum.

THE CONTRIBUTORS

John Baird is Senior Lecturer in Education at the Department of Learning, Assessment and Special Education, Faculty of Education, University of Melbourne. He has an extensive background in teaching in biology education at the tertiary level and has for many years collaborated closely with practicing schoolteachers at both the primary and the secondary level of classroom teaching and research. He was awarded his

Ph.D. in 1984 for a study of a classroom intervention to enhance students' metacognition. Currently, his teaching and research interests have broadened from science education to more general perspectives on students' perceptions in learning, reflection and classroom learning behaviors, metacognition, and teacher professional development.

Sharon Bendall received her B.S. and M.S. degrees in physics, completing her last degree in 1983, and subsequently conducting traditional physics research at the IBM T. J. Watson Laboratory for over two years. During the last eight years she has taught in the Physics Department at San Diego State University and served as a research member of the SDSU Center for Research in Mathematics and Science Education. At SDSU, Bendall has conducted research in physics education under the auspices of several major grants. The results from these projects have been disseminated through numerous publications and presentations. Her recent interests include developing strategies that use writing as a learning tool in the physics classroom and the use of technology in the physics classroom.

Ruth Ben-Zvi is the head of the Chemistry Group in the Department of Science Teaching at the Weizmann Institute of Science in Rehovot, Israel. Since completing her Ph.D. in biochemistry from the Hebrew University of Jerusalem in 1965, she has been involved in chemistry curriculum development. Her research interests are in learning difficulties in chemistry, assessment of students' achievement in chemistry, and affective variables that influence students' learning.

Malcolm Carr is Pro Vice-Chancellor (Academic) and an associate professor in the Centre for Science, Mathematics and Technology Education Research (CSMTER) at the University of Waikato, Hamilton, New Zealand. He earned his Ph.D. in 1962 in chemistry, which he taught at the tertiary level for 28 years, combining this interest with science education for the last 12 years. His research interests are science education related to teaching and learning in the classroom, and more recently technology education with the same focus.

Jere Confrey is an associate professor of mathematics education at the Department of Education at Cornell University, having earlier earned her Ph.D. at Cornell in 1980. She has investigated students' conceptions of mathematics for a number of years, and more recently has been designing software and multimedia products on functions to promote deeper student understanding. She started a math program for women, SummerMath, at Mount Holyoke College. She has just completed a three-year longitudinal study of 20 third- to fifth-graders, completing a teaching experiment using a unique approach to ratio, proportion, multiplication, and splitting. She is Vice President of the International Group for the Psychology of Mathematics Education.

Beatriz S. D'Ambrosio is an associate professor of mathematics education at Indiana University–Purdue University at Indianapolis, having obtained her Ph.D. from Indiana University in 1987. Prior to this appointment, she was a faculty member in the Department of Mathematics Education at the University of Georgia. Her research interests lie in understanding the relationships between research and practice in mathematics education. In particular, she has focused on building preservice and inservice experiences that develop the teacher-researcher, with the intent of empowering teachers and bridging the gap between research and practice.

Helen M. Doerr is Assistant Professor of Mathematics Education in the School of Education and Department of Mathematics at Syracuse University. She earned her Ph.D. from Cornell University in 1994, where she was Associate Director for Scientific Computational Support at the Cornell Theory Center. Her current research interests are in understanding student's learning through model building and the use of computer technology for mathematical problem solving.

Rosalind Driver is Professor of Science Education at King's College at the University of London. Prior to this appointment, she held various positions, most recently, that of Professor of Science Education at the University of Leeds, where for the past 12 years she has been the director of the Children' Learning in Science Research Group. Her long-standing interest in children's ideas about natural phenomena was stimulated during her experience as a physics teacher in Africa and South America, after which she completed her Ph.D. at the Univesity of Illinois. Her professional interests are in the learning of science, teacher development, and the public understanding of science.

Gaalen Erickson is a professor in the Department of Mathematics and Science Education at the University of British Columbia, where he has worked with preservice teachers and graduate students since 1975. Prior to that, he taught science at the high-school level in England and Alberta. He received his Ed.D. in 1975 from the University of British Columbia. His professional concerns include the analysis of how pupils learn more about their world, how beginning teachers learn how to teach, and investigating different models for creating "communities of inquiry" among pupils and teachers.

James J. Gallagher is Professor of Science Education at Michigan State University, where he has worked for the past 19 years. He held university appointments at Stanford University, the Educational Research Council of America, and Governors State University after completing his doctoral degree at Harvard University in 1965. His major research interests are in the ecology of secondary school science classrooms and in the education of science teachers. He currently codirects a project on classroom-based assessment in middle school science and mathematics. He has a strong interest in international science education and currently is part of a Ministry of Education project in Thailand on environmental education. He has led development of interpretive research in science education in several countries, including Taiwan, Brazil, and Panama.

Fred Goldberg is Professor of Physics at San Diego State University and leader of the Physics Learning Research Group of the Center for Research in Mathematics and Science Education. He received his Ph.D. in physics from the University of Michigan in 1971. During the past 15 years he has been involved in investigating student learning in physics and in developing both high-tech and low-tech strategies to promote learning. Currently, he directs a large national project in the United States to develop powerful computer software that would support a constructivist-oriented learning environment in secondary school physics classrooms.

Diane J. Grayson is a physicist and senior lecturer in the Faculty of Science at the University of Natal–Pietermaritzburg, South Africa. Previously, she taught physics at the Durban and Pietermaritzburg campuses of the University of Natal, and completed her doctorate at the University of Washington in 1990. In 1990 she implemented the Science Foundation Programme, designed to provide poorly prepared science students with access to science studies. Similar programs are now being instituted throughout South African universities. Her research interests include student understanding of physics concepts and applying research outcomes to curriculum development and classroom practice.

Richard Gunstone is Professor of Science and Technology Education at Monash University, Melbourne, Australia. Before joining the Monash faculty in 1974, he taught physics and mathematics in Victoria, Alberta, schools, and was awarded his Ph.D. in 1981. Currently, his teaching and research embrace science education, metacognition, and professional development of preservice and in-service teachers. His specific research interests focus on learning and assessment.

Brian Hand is a lecturer in science education at La Trobe University–Bendigo in Victoria, Australia, where his major responsibilities involve teaching postgraduate courses. He was awarded his Ph.D. in 1992 for research involving a long-term evaluation of constructivist teaching within a science department in one school. Prior to taking up his current appointment, he taught science in the secondary schools of several Australian states. His research interests include in-service and preservice teacher education, with an emphasis on implementing constructivist approaches to classroom practice.

John Happs is an educational consultant and is Director of Learning

Performance Seminars, an agency that provides seminars in study techniques for primary, secondary, and tertiary students. He was awarded his D.Phil. from the University of Waikato, New Zealand, in 1983, and was a lecturer in science education at Edith Cowan University and senior lecturer at Murdoch University, both in Perth, Australia, prior to moving into the private sector. His current research interests are in student learning outcomes and curriculum development.

Peter W. Hewson is Professor of Science Education at the University of Wisconsin–Madison, a position he has held since 1985. He received his D.Phil. in theoretical physics in 1969 from Oxford University. During 1971–1984 he held appointments in physics at the University of the Witwatersrand in South Africa. Major influences on his thinking occurred during a postdoctoral fellowship at the University of British Columbia (1969–1971) and a sabbatical at Cornell University (1978–1979). His major research interests are in conceptual change and preservice and in-service teacher education.

Avi Hofstein is a member of the Chemistry Group in the Department of Science Teaching at the Weizmann Institute of Science in Rehovot, Israel, where he earned his Ph.D. for research on the assessment of laboratory work in 1975. His responsibilities in chemistry education extend beyond the Weizmann Institute, and he jointly holds the position of coordinating inspector for high school chemistry. His main research interests are in chemistry education, the affective domain, learning and assessement of laboratory work in chemistry, and science technology and society issues.

Helen Mansfield is Associate Professor of Mathematics Education in the Curriculum Research and Development Group at the University of Hawaii. She received her Ed.D. from the University of Georgia in 1984 for research on students' geometric ideas. Her research interests are in the teaching and learning of geometry; mathematics curricula for the middle grades; and student, teacher, and subject-matter interactions in mathematics classrooms.

Hans Niedderer is Professor of Physics Education in the Institute for Physics Education at the University of Bremen in Germany. He previously held an appointment at the Institute for Science Education (IPN) at the University of Kiel, where he finished his Ph.D. in 1972 with an empirical study on students' conceptions using simple electric circuits. His current research interests include students' learning processes in physics; a new teaching approach in high school quantum physics, including students' understanding of quantum theory and learning processes; computers in physics education; and history and philosophy of physics in physics teaching.

Jeff Northfield is Professor of Education at Monash University in

Melbourne, Australia. Prior to joining the Monash faculty in 1973, he taught in each of the science disciplines in a range of secondary schools, and completed his Ph.D. in 1980. His major research interests are in teacher education, science education, and curriculum research and evaluation. In 1984 and 1993 he returned to secondary school teaching to seek a teacher perspective on implementing teaching and learning ideas in the classroom and implications for teacher education.

Joseph D. Novak is Professor of Education and Professor of Biological Sciences at Cornell University, where he has taught and done research for 28 years. He received his Ph.D. from the University of Minnesota in 1958. Currently, his research centers on the use of metacognitive tools to facilitate new knowledge production in corporate, as well as academic, settings.

Horst P. Schecker holds the position of Research Fellow in Physics Education in the Institute for Physics Education at the University of Bremen in Germany. He obtained his Ph.D. in 1985 with a dissertation on "Students' Alternative Conceptions in Newtonian Mechanics." He has directed two federal R&D projects on computers in science education and taught physics in German high schools. His postdoctoral thesis for qualification as a university professor is about system dynamics modeling in physics education. His current research interests are in the effects of computer-based modeling on physics learning processes and in interactive multimedia.

Philip H. Scott is Lecturer in Science Education at the University of Leeds. Prior to taking up this position in 1994, he taught physics and science in Leeds high schools and became Co-ordinator of the Children's Learning in Science Research Group in 1990. He is interested in the ways in which findings from research into children's learning of science concepts might be drawn upon to inform science teaching. Currently he is investigating how language mediates science learning in the classroom.

Leslie P. Steffe is Research Professor of Mathematics Education in the Department of Mathematics Education at the University of Georgia. He earned his Ph.D. from the University of Wisconsin in 1966, where he spent a year in postdoctoral study prior to joining the faculty of mathematics education at Georgia. His research interests are in children's construction of mathematical concepts and operations and the development of a constructivist orientation in mathematics teaching and learning.

Kenneth Tobin has been Professor of Science Education at Florida State University since 1988. Prior to that, he taught science education at Curtin University and Edith Cowan University and taught physics and science in several schools in Western Australia. He earned his doctorate from the University of Georgia in 1980. He has pioneered the use of qualitative research in science education and has a major research interest in educational reform.

Richard White is Dean of Education at Monash University. Prior to becoming Dean in 1994, he was Professor of Educational Psychology. Before joining the Monash faculty in 1971, he taught science in secondary schools. His Ph.D. was awarded in 1972, for research into the learning of kinematics. His subsequent research continued the study of learning, especially of science.

Index